SRA
Connecting Math Concepts

Level C Teacher's Guide

COMPREHENSIVE EDITION

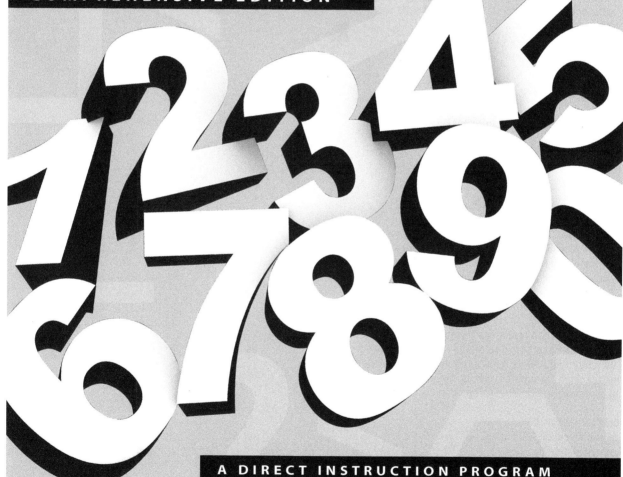

A DIRECT INSTRUCTION PROGRAM

McGraw Hill Education

Bothell, WA • Chicago, IL • Columbus, OH • New York, NY

Acknowledgments

The authors are grateful to the following people for their input in the field-testing and preparation of *SRA Connecting Math Concepts: Comprehensive Edition Level C:*

Jennica Braxton
Amilcar Cifuentes
Don Crawford
Ashly Cupit
Jennifer Ditty
Laura Doherty
Karen Fierman
Judy Hein
Jeff Kinsella
Debbi Kleppen
Melissa Kramer
Margie Mayo
Richelle Owen
Toni Reeves
Mary Rosenbaum
Harry Schilling
Jerry Silbert
Alicia Smith
Leigh VonDerahe
Jason Yanok

Photo Credits

85 (l to r, t to b)Comstock Images/Alamy, (2) Comstock Images/Alamy, (3)Angelika Antl/age fotostock, (4) RubberBall Productions/Getty Images, (5) Siede Preis/Photodisc/Getty Images, (6) Howard Shooter/Dorling Kindersley RF/Getty Images, (7) Jonathan Kantor/Getty Images, (8) Mark Steinmetz/McGraw-Hill Education, (9 & 10) Steven Puetzer/Photographer's Choice/Getty Images; **131** (l) Walter Quirtmair/Panther Media/age fotostock, (r) Andriy Popov/123RF.com; **164** (l) Siede Preis/Photodisc/Getty Images, (cl)McGraw-Hill Education, (cr) Howard Shooter/Dorling Kindersley RF/Getty Images, (r) Mark Steinmetz/McGraw-Hill Education; **168** (l) Siede Preis/Photodisc/Getty Images, (cl)McGraw-Hill Education, (cr) Howard Shooter/Dorling Kindersley RF/Getty Images, (r) Mark Steinmetz/McGraw-Hill Education

MHEonline.com

Send all inquiries to:
McGraw-Hill Education
4400 Easton Commons
Columbus, OH 43219

ISBN: 978-0-02-103594-6
MHID: 0-02-103594-6

Printed in the United States of America.

16 17 LKV 25 24 23

Contents

Program Overview

The Connecting Math Concepts: Comprehensive Edition

Connecting Math Concepts: Comprehensive Edition is a six-level series that will accelerate the math learning performance of students in grades K through 5. Levels A through F are suitable for regular-education students in Kindergarten through fifth grade. The series is also highly effective with at-risk students.

Connecting Math Concepts: Comprehensive Edition is based on the fact that understanding mathematics requires making connections

- among related topics in mathematics, and
- between procedures and knowledge.

Connecting Math Concepts: Comprehensive Edition does more than expose students to connections. It stresses understanding and introduces concepts carefully, then weaves them together throughout the program. Once something is introduced, it never goes away. It often becomes a component part of an operation that has several steps. After appearing for several lessons in structured exercises, strategies and concepts are systematically reviewed as Independent Work.

The organization of *Connecting Math Concepts: Comprehensive Edition* is powerful because lessons have been designed to

- Teach explicit strategies that all students can learn and apply.
- Introduce concepts at a reasonable rate, so all students make steady progress.
- Help students make connections between important concepts.
- Provide the practice needed to achieve mastery and understanding.
- Meet the math standards specified in the Common Core State Standards for Mathematics.

The program's Direct Instruction design permits significant acceleration of student performance. The instructional sequences are the same for all students, but the rate at which students proceed through each level should be adjusted according to student performance. Higher performers proceed through the levels faster. Lower performers receive more practice. Benchmark in-program Mastery Tests provide information about how well students are mastering what has been taught most recently. Students' daily performance and test performance disclose whether they need more practice or whether they are mastering the material on the current schedule of lesson introduction.

The program enables the teacher to teach students at a faster rate and with greater understanding than they probably ever achieved before. The scripted lessons have been shaped through extensive field-testing and classroom observation. The teacher individualizes instruction to accommodate different groups that make different mistakes and require different amounts of practice to learn the material.

Introduction to 2012 CMC Level C

- *CMC Level C* is designed for students who have successfully completed *CMC Level B* or who pass the Placement Test for *Level C* (see page 135).
- *CMC Level C* instruction meets all requirements of the Common Core State Standards for Mathematics for second grade.

Program Information

The following summary table lists facts about
2012 *CMC Level C*.

Students who are appropriately placed in *CMC Level C*	Pass Placement Test (p. 135)
How students are grouped	Instructional groups should be as homogeneous as possible.
Number of lessons	• 130 regular lessons • 13 Mastery Tests • 2 Cumulative Tests
Schedule	• 50 minutes for structured work • Additional 20 minutes for students' Independent Work • 5 periods per week

Teacher Material	• Teacher's Guide
	• Presentation Book 1:
	    Lessons 1–70, Mastery Tests 1–7
	    Cumulative Test 1 (Lessons 1–70)
	• Presentation Book 2:
	    Lessons 71–130, Mastery Tests 8–13
	    Cumulative Test 2 (Lessons 1–130)
	• Answer Key
	• Board Displays

Student Material	• Workbook 1: Lessons 1–70
	• Workbook 2: Lessons 71–130
	• Textbook
	• Student Assessment Book:
	    Mastery Tests 1–13, Cumulative Tests 1 and 2
	    Remedies worksheets for Mastery Tests 1–13, Cumulative Tests 1 and 2

In-Program Tests	13 ten-lesson Mastery Tests • Administration and Remedies are specified in the Teacher Presentation Books. • Tests and Remedies worksheets are in the Student Assessment Book.
Optional Cumulative Tests	2 Cumulative Tests • Administration is specified in Teacher Presentation Books 1 and 2. • Test sheets and Remedies worksheets are in the Student Assessment Book.
Additional Teacher/ Student Material	• Math Fact Worksheets (Online Blackline Masters via ConnectED) • Access to *CMC* content online via ConnectED • *SRA 2Inform* available on ConnectED for online progress monitoring

TEACHER MATERIAL

The teacher material consists of:

The Teacher's Guide: This guide explains the program and how to teach it properly. The Scope and Sequence chart on pages 10–11 shows the various tracks (topics or strands) that are taught;

indicates the starting lesson for each track/ strand; and shows the lesson range. This guide calls attention to potential problems and provides information about how to present exercises and how to correct specific mistakes the students may make. The guide is designed to be used to help you teach more effectively.

Two Teacher Presentation Books: These books specify each exercise in the lessons and tests to be presented to the students. The exercises provide scripts that indicate what you are to say, what you are to do, the responses students are to make, and correction procedures for common errors. (See Teaching Effectively, **Using the Teacher Presentation Scripts,** for details about using the scripts.)

Answer Key: The answers to all of the problems, activities, and tests appear in the Answer Key to assist you in checking the students' class work and Independent Work and for marking tests. The Answer Key also specifies the remedy exercises for each test and provides a group summary of test performance.

Board Displays: The teacher materials include the Board Displays, which show all the displays you present during the lessons. This component is flexible and can be utilized in different ways to support the instruction—via a computer hooked up to a projector, to a television, or to any interactive white board. The electronic Board Displays are available online on ConnectED. You can navigate through the displays with a touch of the finger if you have an interactive white board or with the click of a button from a mouse (wired or wireless) or a remote control.

ConnectED: On McGraw-Hill/SEG's ConnectED platform you can plan and review *CMC* lessons and see correlations to Common Core State Standards for Mathematics. Access the following *CMC* materials from anywhere you have an Internet connection: PDFs of the Presentation Books, an online planner, online Board Displays, Math Fact Worksheets, eBooks of the Teacher's Guides, and correlations. *CMC* on ConnectED also features a progress monitoring application called *SRA 2Inform* that stores student data and provides useful reports and graphs about student progress. Refer to the card you received with your teacher materials kit for more information about redeeming your access code, good for one six-year teacher subscription and 10 student seat licenses, which provide access to the eTextbook.

STUDENT MATERIAL

The student materials include a set of two Workbooks for each student, a Textbook, and a Student Assessment Book. The Textbook and Workbooks contain writing activities, which the students do as part of the structured presentation of a lesson and as independent seatwork. The Student Assessment Book contains material for the Mastery Tests as well as test Remedies worksheet pages and optional Cumulative Test pages.

Textbook: Lessons 41–130

Workbook 1: Lessons 1–70

Workbook 2: Lessons 71–130

Student Assessment Book: Mastery Tests 1–13, Cumulative Tests 1 and 2, and test Remedies worksheets for Mastery Tests

WHAT'S NEW IN 2012 *CMC LEVEL C*

Most instructional strategies are the same as those of the earlier *CMC* editions; however, the procedures for teaching these strategies have been greatly modified to address problems teachers had teaching the content of the previous editions to at-risk students. The 2012 edition of *CMC Level C* has also been revised on the basis of field-testing.

- The 2012 edition provides far more oral work than earlier editions. This work is presented as "hot series" of tasks. The series are designed so that students respond to ten or more related questions or directions per minute; therefore, these series present a great deal of information about an operation or discrimination in a short period of time.

- The content is revised so that students learn not only the basics but also the higher-order concepts. The result is that second-grade students who complete *CMC Level C* are able to work the full range of problems and applications that define understanding of second-grade math.

- The hallmark of Direct Instruction mathematics programs is that they teach all the component skills and operations required to provide a solid foundation in topics involving place value and operations, money, geometry, measurement and data, and word problems. The *CMC Level C* program addresses all standards specified in the Common Core State Standards for Mathematics for second grade math. (See pages 12–13 and 125–133.)

- *CMC Level C* has support/enhancements, including technology components, for teachers and students. These enhancements include displays in the Teacher Presentation Books, Board Displays online, Workbook and Answer Key pages reduced in the Teacher Presentation Books, a Student Assessment Book with all program assessments in one location, *SRA 2Inform* for online progress monitoring, and the ability to plan and review lessons online via ConnectED.

The Structure of Connecting Math Concepts Level C

Connecting Math Concepts Level C is appropriate for students who complete *Level B* or who pass the *Level C* Placement Test.

CMC Level C has two starting points: Lesson 1 and Lesson 11. Lessons 1–10 are designed to acquaint new students with the conventions that continuing students learned in *Level B*. Continuing students start at Lesson 11. Lessons are designed to review facts and other information presented in *Level B* and to introduce new material at a rate that would be appropriate for both the continuing students and the students who started at Lesson 1.

If you have a group that has new and continuing students, the simplest solution is to start all students at Lesson 1 and proceed as quickly as the lower performers, who are appropriately placed in the program, are able to proceed.

Reproducible copies of the Placement Test appear on pages 139–141 of this guide.

SCHEDULING

The program contains 130 lessons and 13 in-program Mastery Tests. The ideal goal is to teach one lesson each period. If students are not firm on content that is being introduced, you will need to repeat parts of lessons or entire lessons. Particularly early in the program, you may need to repeat entire lessons because students will perform much better on subsequent lessons if all lessons are taught to mastery.

Also, some lessons are longer and may require more than a period to complete. Following long lessons, try to get back on a schedule of teaching a lesson a day. This pattern assures that students receive daily practice in skills or operations that have been recently introduced.

The program is to be taught daily. Periods for structured work are 50 minutes. Students need an additional 20 minutes or more to complete their Independent Work. If the Independent Work cannot be completed in school, it may be assigned as homework, but this is not an attractive alternative, particularly on Fridays. It's important for students to bring back their work on the following school day. If you assign Textbook Independent Work as homework, the Textbooks should remain in the classroom.

Two Teacher Presentation Books: These books specify each exercise in the lessons and tests to be presented to the students. The exercises provide scripts that indicate what you are to say, what you are to do, the responses students are to make, and correction procedures for common errors. (See Teaching Effectively, **Using the Teacher Presentation Scripts,** for details about using the scripts.)

Answer Key: The answers to all of the problems, activities, and tests appear in the Answer Key to assist you in checking the students' class work and Independent Work and for marking tests. The Answer Key also specifies the remedy exercises for each test and provides a group summary of test performance.

Board Displays CD: The teacher materials include the Board Displays CD, which shows all the displays you present during the lessons. This component is flexible and can be utilized in different ways to support the instruction— via a computer hooked up to a projector, to a television, or to any interactive white board. The electronic Board Displays are also available online on ConnectED. You can navigate through the displays with a touch of the finger if you have an interactive white board or with the click of a button from a mouse (wired or wireless) or a remote control.

Practice Software: The *CMC Level C* Practice Software provides students additional practice with the skills and concepts taught in *CMC Level C*. It is a core component for meeting several Common Core State Standards for Mathematics. Students apply their skills to tasks presented onscreen. The tasks are governed by an algorithm that adjusts the amount of practice students receive according to how well they perform. Games and reward screens provide students with reinforcement for meeting performance goals. The software is organized into blocks, each presenting activities for a 20-lesson segment of the program as students proceed through the lessons.

The Math Facts strand of the software is organized into sets of facts that follow the instructional sequence in the lessons. It is designed to facilitate continuous review and reinforcement of the math facts as they are introduced and practiced. It is available via ConnectED with 10 student seat licenses per every teacher materials kit purchase.

ConnectED: On McGraw-Hill/SEG's ConnectED platform you can plan and review *CMC* lessons and see correlations to Common Core State Standards for Mathematics. Access the following *CMC* materials from anywhere you have an Internet connection: PDFs of the Presentation Books, an online planner, online printable versions of the Board Displays CD, student Practice Software (online version requires separately purchased student licenses), Math Fact Worksheets, eBooks of the Teacher's Guides, and correlations. *CMC* on ConnectED also features a progress monitoring application called *SRA 2Inform* that stores student data and provides useful reports and graphs about student progress. Refer to the card you received with your teacher materials kit for more information about redeeming your access code, good for one six-year teacher subscription and 10 student seat licenses, which provide access to the Practice Software and eTextbook.

STUDENT MATERIAL

The student materials include a set of two Workbooks for each student, a Textbook, and a Student Assessment Book. The Textbook and Workbooks contain writing activities, which the students do as part of the structured presentation of a lesson and as independent seatwork. The Student Assessment Book contains material for the Mastery Tests as well as test Remedies worksheet pages and optional Cumulative Test pages.

Textbook: Lessons 41–130

Workbook 1: Lessons 1–70

Workbook 2: Lessons 71–130

Student Assessment Book: Mastery Tests 1–13, Cumulative Tests 1 and 2, and test Remedies worksheets for Mastery Tests

What's New in 2012 *CMC Level C*

Most instructional strategies are the same as those of the earlier *CMC* editions; however, the procedures for teaching these strategies have been greatly modified to address problems teachers had teaching the content of the previous editions to at-risk students. The 2012 edition of *CMC Level C* has also been revised on the basis of field-testing.

- The 2012 edition provides far more oral work than earlier editions. This work is presented as "hot series" of tasks. The series are designed so that students respond to ten or more related questions or directions per minute; therefore, these series present a great deal of information about an operation or discrimination in a short period of time.

- The content is revised so that students learn not only the basics but also the higher-order concepts. The result is that second-grade students who complete *CMC Level C* are able to work the full range of problems and applications that define understanding of second-grade math.

- The hallmark of Direct Instruction mathematics programs is that they teach all the component skills and operations required to provide a solid foundation in topics involving place value and operations, money, geometry, measurement and data, and word problems. The *CMC Level C* program addresses all standards specified in the Common Core State Standards for Mathematics for second grade math. (See pages 12–13 and 125–133.)

- *CMC Level C* has support/enhancements, including technology components, for teachers and students. These enhancements include displays in the Teacher Presentation Books, a Board Displays CD (also available online), Workbook and Answer Key pages reduced in the Teacher Presentation Books, a Student Assessment Book with all program assessments in one location, *SRA 2Inform* for online progress monitoring, student Practice Software, and the ability to plan and review lessons online via ConnectED.

The Structure of Connecting Math Concepts Level C

Connecting Math Concepts Level C is appropriate for students who complete *Level B* or who pass the *Level C* Placement Test.

CMC Level C has two starting points: Lesson 1 and Lesson 11. Lessons 1–10 are designed to acquaint new students with the conventions that continuing students learned in *Level B*. Continuing students start at Lesson 11. Lessons are designed to review facts and other information presented in *Level B* and to introduce new material at a rate that would be appropriate for both the continuing students and the students who started at Lesson 1.

If you have a group that has new and continuing students, the simplest solution is to start all students at Lesson 1 and proceed as quickly as the lower performers, who are appropriately placed in the program, are able to proceed.

Reproducible copies of the Placement Test appear on pages 139–141 of this guide.

SCHEDULING

The program contains 130 lessons and 13 in-program Mastery Tests. The ideal goal is to teach one lesson each period. If students are not firm on content that is being introduced, you will need to repeat parts of lessons or entire lessons. Particularly early in the program, you may need to repeat entire lessons because students will perform much better on subsequent lessons if all lessons are taught to mastery.

Also, some lessons are longer and may require more than a period to complete. Following long lessons, try to get back on a schedule of teaching a lesson a day. This pattern assures that students receive daily practice in skills or operations that have been recently introduced.

The program is to be taught daily. Periods for structured work are 50 minutes. Students need an additional 20 minutes or more to complete their Independent Work. If the Independent Work cannot be completed in school, it may be assigned as homework, but this is not an attractive alternative, particularly on Fridays. It's important for students to bring back their work on the following school day. If you assign Textbook Independent Work as homework, the Textbooks should remain in the classroom.

and subtracting. Later, these skills are combined to work problems with letters (not words), such as: 18 is 12 less than R, and P is 14 less than 56. Students make the families with two numbers and a letter. They solve for the letter.

After all these skills are well practiced, word problems are introduced.

Scripted Presentations

All exercises in each lesson are scripted. The script indicates the wording you use in presenting the material and correcting student errors. Once you are familiar with the program, you may deviate some from the exact wording; however, until you know why things are phrased as they are, you should follow the exact wording. The most common mistakes teachers make in presenting the material is to rephrase some instructions. Later, when the original instructions become components of more complicated operations, the students are not prepared to respond to steps that have variant wording.

In *Connecting Math Concepts Level C,* you first present material in a structured sequence that requires students to respond verbally. This technique permits you to present tasks at a high rate so it is very efficient for teaching. Also, it provides you with information about which students are responding correctly and which need more repetition.

Typically, after students respond to a series of verbal tasks, you present written work.

Here's an exercise with a verbal series in which students identify whether the big number or a small number is missing, and, therefore, whether they add or subtract to find the missing number.

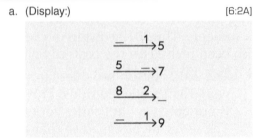

a. (Display:)   [6:2A]

A number is missing in each family. You'll say the problems.
- If the big number is missing, do you add or subtract? (Signal.) *Add.*
- What do you do if a small number is missing? (Signal.) *Subtract.*
b. (Point to ═══¹→5.) Is the big number missing in this family? (Signal.) *No.*
- So what do you do to find the missing number? (Signal.) *Subtract.*
- Say the problem. (Signal.) *5 – 1.*
c. (Point to ⁵═══→7.) Is the big number missing in this family? (Signal.) *No.*
- So what do you do to find the missing number? (Signal.) *Subtract.*
- Say the problem. (Signal.) *7 – 5.*
d. (Point to ⁸══²→_.) Is the big number missing in this family? (Signal.) *Yes.*
- So what do you do to find the missing number? (Signal.) *Add.*
- Say the problem. (Signal.) *8 + 2.*
e. (Point to ═══¹→9.) Is the big number missing in this family? (Signal.) *No.*
- So what do you do to find the missing number? (Signal.) *Subtract.*
- Say the problem. (Signal.) *9 – 1.*
(Repeat until firm.)

A great advantage of sequences like these is that if students later make mistakes involving the discriminations the series addressed, you have information about exactly how to correct the mistakes. For instance, if students make mistakes of doing the wrong operation, your correction would be to ask:

- Is the big number missing?
- So what do you do to find the missing number?

The scripted presentation is designed to help you present the key discriminations quickly and with consistent language, which helps maximize the efficiency of your teaching.

Language

Connecting Math Concepts Level C does not always initially use the traditional mathematical vocabulary associated with some content. The reason is that what is being taught occurs in stages over many lessons—not all at once in a single lesson or several lessons. The language students need to solve traditional problems will ultimately be taught. For example, finding the number of squares in a rectangle is introduced in Lesson 44. In Lesson 96, square units (square inches, square centimeters, square feet) are introduced. The term *area* is introduced on Lesson 97.

The general format of introduction calls for a minimum of vocabulary and a strong emphasis on demonstrating how the operation works, what the discriminations are, and which steps are needed to solve problems. Vocabulary that is not essential to solving a problem type will probably not be introduced.

Scope and Sequence for Connecting Math Concepts Level C

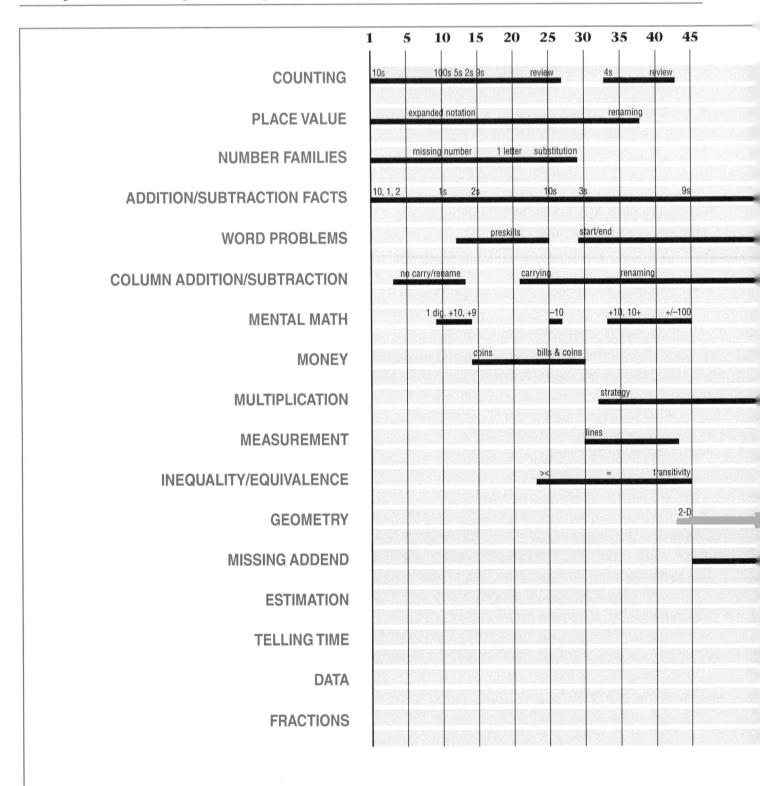

	1	5	10	15	20	25	30	35	40	45
COUNTING	10s		100s 5s 2s 9s			review		4s	review	
PLACE VALUE		expanded notation					renaming			
NUMBER FAMILIES		missing number		1 letter	substitution					
ADDITION/SUBTRACTION FACTS	10, 1, 2	1s	2s			10s	3s		9s	
WORD PROBLEMS				preskills		start/end				
COLUMN ADDITION/SUBTRACTION	no carry/rename				carrying		renaming			
MENTAL MATH		1 dig. +10, +9			–10		+10, 10+	+/–100		
MONEY		coins		bills & coins						
MULTIPLICATION						strategy				
MEASUREMENT					lines					
INEQUALITY/EQUIVALENCE					><	=	transitivity			
GEOMETRY								2-D		
MISSING ADDEND										
ESTIMATION										
TELLING TIME										
DATA										
FRACTIONS										

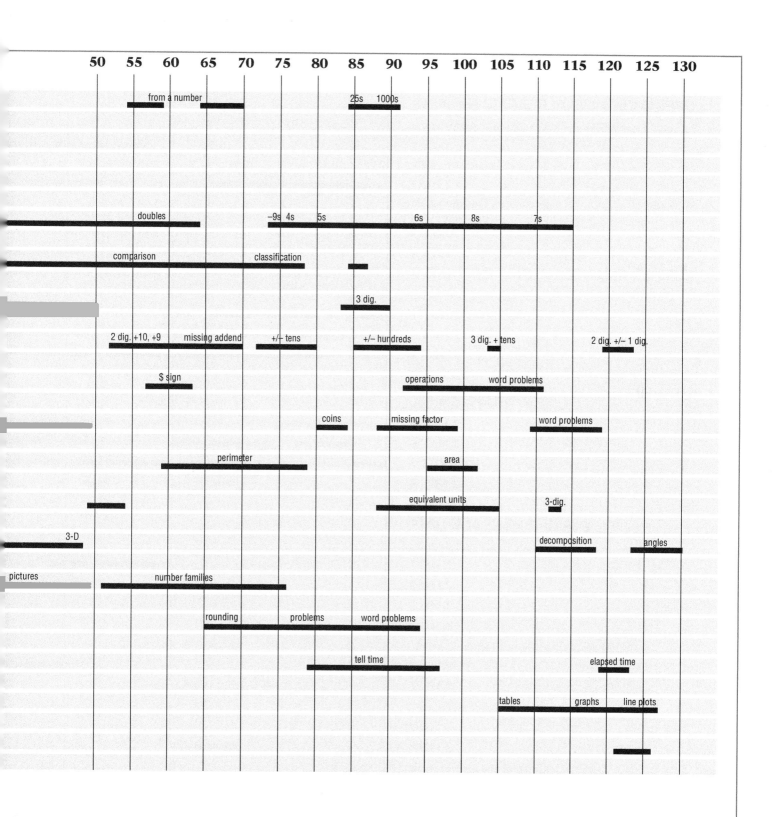

50	55	60	65	70	75	80	85	90	95	100	105	110	115	120	125	130

from a number

25s 1000s

doubles −9s 4s 5s 6s 8s 7s

comparison classification

3 dig.

2 dig. +10, +9 missing addend +/− tens +/− hundreds 3 dig. + tens 2 dig. +/− 1 dig.

$ sign operations word problems

coins missing factor word problems

perimeter area

equivalent units 3-dig.

3-D decomposition angles

pictures number families

rounding problems word problems

tell time elapsed time

tables graphs line plots

Common Core State Standards Chart and *CMC Level C*

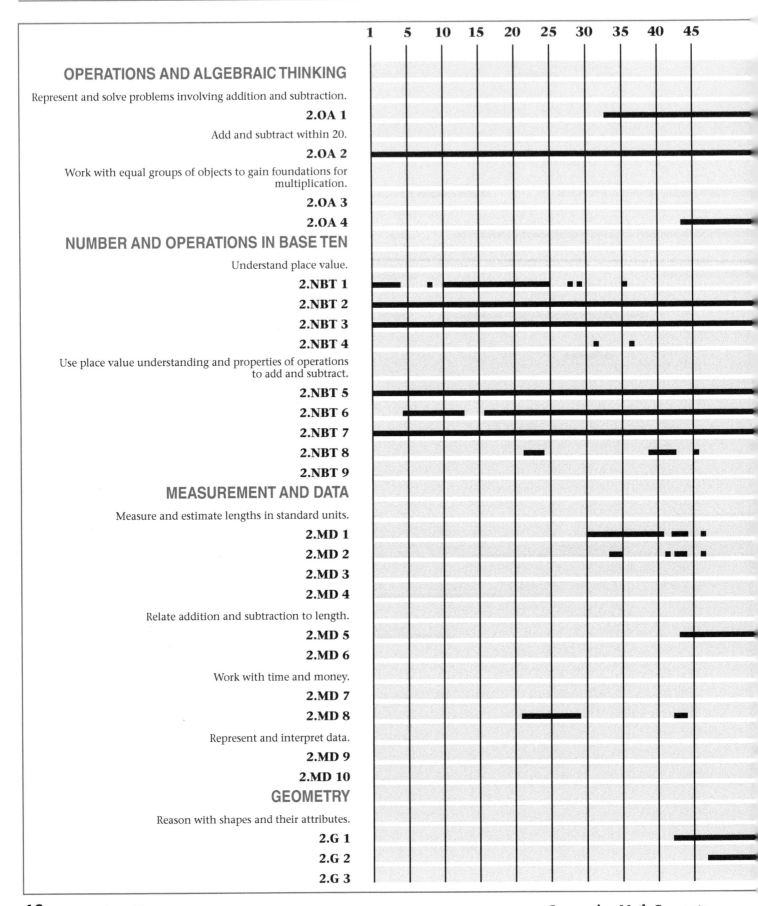

	1	5	10	15	20	25	30	35	40	45

OPERATIONS AND ALGEBRAIC THINKING

Represent and solve problems involving addition and subtraction.

2.OA 1

Add and subtract within 20.

2.OA 2

Work with equal groups of objects to gain foundations for multiplication.

2.OA 3

2.OA 4

NUMBER AND OPERATIONS IN BASE TEN

Understand place value.

2.NBT 1

2.NBT 2

2.NBT 3

2.NBT 4

Use place value understanding and properties of operations to add and subtract.

2.NBT 5

2.NBT 6

2.NBT 7

2.NBT 8

2.NBT 9

MEASUREMENT AND DATA

Measure and estimate lengths in standard units.

2.MD 1

2.MD 2

2.MD 3

2.MD 4

Relate addition and subtraction to length.

2.MD 5

2.MD 6

Work with time and money.

2.MD 7

2.MD 8

Represent and interpret data.

2.MD 9

2.MD 10

GEOMETRY

Reason with shapes and their attributes.

2.G 1

2.G 2

2.G 3

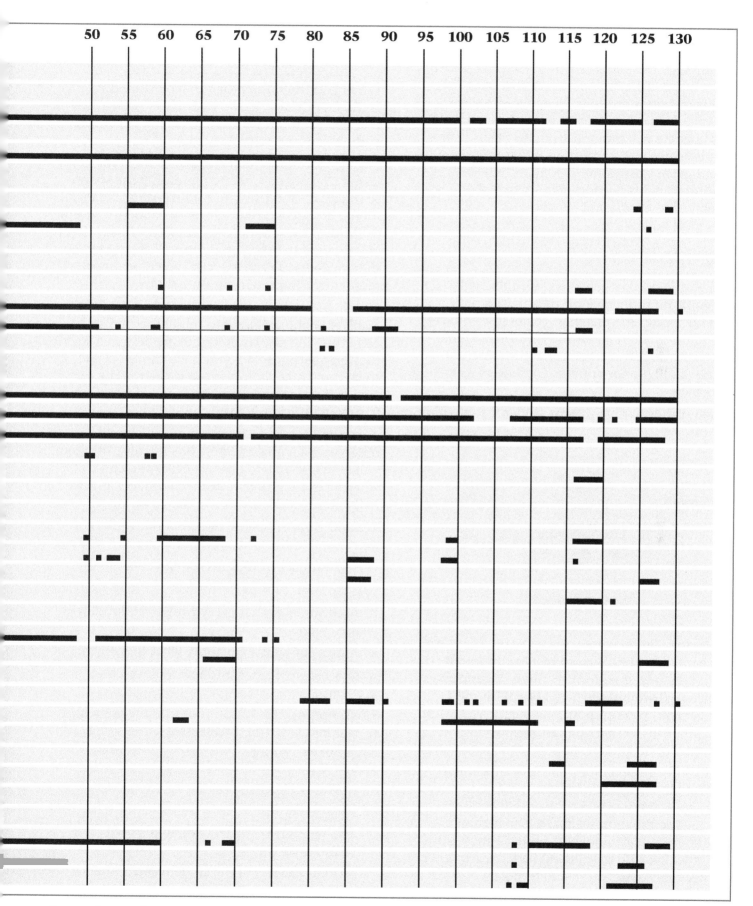

Teaching Effectively

Connecting Math Concepts Level C is designed for students who have the necessary entry skills measured by the Placement Test. (See Placement Tests on page 135.) The group should be as homogeneous as possible. Students who have similar entry skills and learn at approximately the same rate will progress through the program more efficiently as a group. So if there are three second-grade classrooms, it could be efficient to group students homogeneously (based on placement-test scores).

Organization

Even within a homogeneous class, there will be significant differences in the rate at which students master the material. The best way to get timely information about the performance is to arrange seating so you can receive information quickly on higher performers and lower performers.

A good plan is to organize the students something like this:

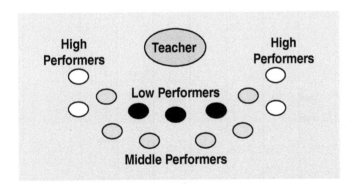

The lowest performers are closest to the front of the classroom. Middle performers are arranged around the lowest performers. Highest performers are arranged around the periphery. With this arrangement, you can position yourself as students work problems so that you can sample low, average, and high performers by taking a few steps.

While different variations of this arrangement are possible, be careful not to seat low performers far from the front center of the room because they require the most feedback. The highest performers, understandably, can be farthest from the center because they attend better, learn faster, and need less observation and feedback.

Using the Teacher Presentation Scripts

When you teach the program, you should be familiar with each lesson before you present it so that you can monitor student responses, during both verbal and written exercises.

Ideally, you should rehearse any parts of the lesson that are new before presenting the lesson to the class. Don't simply read the script, but act it out before you present it to the students. Attend to the displays and how the displays change. If you preview the steps students will take to work the problems in each exercise, you'll be much more fluent in presenting the activity.

Watch your wording. Activities that don't involve displays are much easier to present than display activities. The display activities are designed so they are manageable if you have an idea of the steps you'll take. If you rehearse each of the early lessons before presenting them, you'll learn how to present efficiently from the script.

As students work each problem, you should observe an adequate sample of students. Although you won't be able to observe every student working every problem, you can observe at least half a dozen students in less than a minute.

Remind students of the two important rules for doing well in this program:

1. Always work problems the way they are shown.

2. No shortcuts are permitted.

Remind students that everything introduced will be used later.

Reinforce students who apply what they learn.

Always require students to rework incorrect problems.

The script for each lesson indicates precisely how to present each structured activity. The script shows what you say, what you do, and what the students' responses should be.

What you say appears in blue type:

You say this.

What you do appears in parentheses:

(You do this.)

The responses of the students are in italics.

Students say this.

Although you may feel uncomfortable "reading" a script (and you may feel that the students will not pay attention), try to present the exercises as if you're saying something important to the students. If you do, you'll find that working from a script is not difficult and that students respond well to what you say. A sample script appears on page 16 of this guide.

The arrows show five different things you'll do in addition to delivering the wording in the script.

1. You'll **signal** to make sure group responses involve all the students. (arrow 1)

2. You'll **firm** critical parts of the exercises. (arrow 2)

3. You'll **pace** your presentation based on what the students are doing, judging whether to proceed quickly or to wait a few more seconds before moving on with the presentation. (arrow 3)

4. You'll **display** (or write) things on the board, and you'll often **add** to the board display. (arrow 4)

5. You'll **check** students' written work to ensure mastery of the content. (arrow 5)

ARROW 1: GROUP RESPONSES (SIGNAL.)

Some of the tasks call for group responses. If students respond with brisk unison responses, you receive good information about whether most of the students are performing correctly. The simplest way to signal students to respond together is to adopt a timing practice—just like the timing in a musical piece.

A signal follows a question (as shown by arrow 1) or a direction.

You can signal with a hand drop, clapping one time, snapping your fingers, or tapping your foot. After initially establishing the timing for signals, you can signal through voice inflection and timing.

Students will not be able to initiate responses together at the appropriate rate unless you follow these rules:

a. Talk first. Pause a standard length of time (possibly 1 second). Then signal. Never signal while you talk. Don't change the timing from one signal to the next. For tasks in which you point to a problem on the board or a screen,

first point to the item. Talk and then signal. Students are to respond on your signal—not after it or before it—and respond at a natural speaking rate.

b. Model responses at the rate you expect students to produce them. Don't permit students to produce slow, droning responses because droning responses can give students the opportunity to copy the responses of others. If students are required to respond at a reasonable speaking rate, all must initiate responses, and it's relatively easy to determine which students are responding correctly and which are giving the wrong response. Also, don't permit students to respond at a very fast rate or to "jump" your signal. Listen very carefully to the first part of the response.

c. Do not respond with the students unless you are trying to help them with a difficult response. You present only what's in blue. You do not say the answers with the students, and you should not move your lips or give any clues to the answer.

Think of signals this way: If you use them correctly, they provide you with much diagnostic information. A weak response suggests that you should repeat a task. Signals are important, therefore, early in the program. After students have learned the routine, the students will be able to respond on cue with no signal. That will happen, however, only if you always give your signal at the end of a constant time interval after you complete what you say.

ARROW 2: FIRMING (REPEAT UNTIL FIRM.)

When students make mistakes, correct them. A correction may occur during any part of the teacher presentation that calls for the students to respond. It may also occur in connection with what the students are writing.

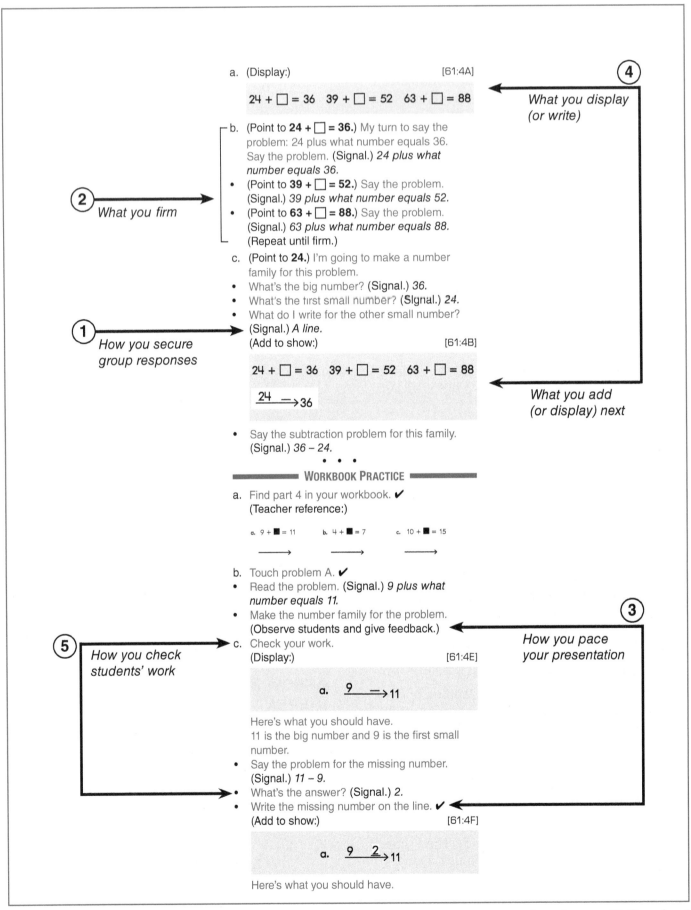

a. (Display:)   [61:4A]

$$24 + \square = 36 \quad 39 + \square = 52 \quad 63 + \square = 88$$

b. (Point to **24 + □ = 36.**) My turn to say the problem: 24 plus what number equals 36. Say the problem. (Signal.) *24 plus what number equals 36.*
• (Point to **39 + □ = 52.**) Say the problem. (Signal.) *39 plus what number equals 52.*
• (Point to **63 + □ = 88.**) Say the problem. (Signal.) *63 plus what number equals 88.* (Repeat until firm.)

c. (Point to **24.**) I'm going to make a number family for this problem.
• What's the big number? (Signal.) *36.*
• What's the first small number? (Signal.) *24.*
• What do I write for the other small number? (Signal.) *A line.* (Add to show:)   [61:4B]

$$24 + \square = 36 \quad 39 + \square = 52 \quad 63 + \square = 88$$

$$\underline{24} \longrightarrow 36$$

• Say the subtraction problem for this family. (Signal.) *36 – 24.*

• • •

WORKBOOK PRACTICE

a. Find part 4 in your workbook. ✔ (Teacher reference:)

a. $9 + \blacksquare = 11$ b. $4 + \blacksquare = 7$ c. $10 + \blacksquare = 15$

$\longrightarrow$ $\longrightarrow$ $\longrightarrow$

b. Touch problem A. ✔
• Read the problem. (Signal.) *9 plus what number equals 11.*
• Make the number family for the problem. (Observe students and give feedback.)

c. Check your work. (Display:)   [61:4E]

a. $\underline{9} \longrightarrow 11$

Here's what you should have. 11 is the big number and 9 is the first small number.
• Say the problem for the missing number. (Signal.) *11 – 9.*
• What's the answer? (Signal.) *2.*
• Write the missing number on the line. ✔ (Add to show:)   [61:4F]

a. $\underline{9 \quad 2} \longrightarrow 11$

Here's what you should have.

② What you firm

① How you secure group responses

④ What you display (or write)

What you add (or display) next

③ How you pace your presentation

⑤ How you check students' work

Here is the rule for correcting mistakes on an oral task: ***Correct the mistake as soon as you hear it.***

A mistake is either saying the wrong thing or not responding.

- To correct:
  1. Say the correct answer.
  2. Repeat the task the students missed.

Here is the first task from step C:

- What's the big number? (Signal.) *36.*

If some students do not respond or say anything other than 36, there's a mistake. As soon as you hear the mistake, correct it.

1. Say the correct answer.
- 36.
2. Repeat the task students missed.
- What's the big number? (Signal.) *36.*

Another type of mistake is responding too quickly or too slowly. To correct oral responses that are too fast or too slow, model exactly how you want students to respond. Here is the correction for the step B Workbook practice:

- My turn to read the problem: 9 plus what number equals 11.
- Your turn: Read the problem. (Signal.) *9 plus what number equals 11.*
- Good saying it the right way.

Remember, wherever there's a signal, there's a place where students may make mistakes. Correct mistakes as soon as you hear them.

A special correction is needed for correcting mistakes on tasks that teach a relationship or include a series of related examples.

This type of correction is marked with a bracket and the note **(Repeat until firm.)**.

The note usually occurs when students must produce more than one related response as in step B of the oral work in the sample script:

b. (Point to **24 +** ☐ **= 36.**) My turn to say the problem:
24 plus what number equals 36.
Say the problem. (Signal.) *24 plus what number equals 36.*

- (Point to **39 +** ☐ **= 52.**) Say the problem. (Signal.)
39 plus what number equals 52.

- (Point to **63 +** ☐ **= 88.**) Say the problem. (Signal.)
63 plus what number equals 88.
(Repeat until firm.)

When you repeat until firm, you follow these steps:
1. Correct the mistake. (Tell the answer and repeat the task that was missed.)
2. Complete the series of bracketed tasks.
3. Return to the beginning of the bracket and repeat the entire series of tasks.

For example, students make a mistake in step B:

- (Point to **39 +** ☐ **= 52.**) Say the problem. (Signal.)
39 plus what number equals 52.

Some students don't respond:

1. **Correct the mistake.**
- 39 plus what number equals 52. Say the problem. (Signal.) *39 plus what number equals 52.*

2. **Complete the series of bracketed tasks.**
- (Point to **63 +** ☐ **= 88.**) Say the problem. (Signal.)
63 plus what number equals 88.

3. **Return to the beginning of the bracket and repeat the entire series of tasks.**
- Let's try that again with no mistakes.

b. (Point to **24 +** ☐ **= 36.**) My turn to say the problem:
24 plus what number equals 36.
Say the problem. (Signal.) *24 plus what number equals 36.*

- (Point to **39 +** ☐ **= 52.**) Say the problem. (Signal.)
39 plus what number equals 52.

- (Point to **63 +** ☐ **= 88.**) Say the problem. (Signal.)
63 plus what number equals 88.
(Repeat until firm.)

Students show you through their responses whether or not the correction worked, whether or not they are firm.

The repeat-until-firm direction usually focuses on knowledge that is very important for later work. As a general procedure, follow the repeat-until-firm directions. Following the repeat-until-firm directions will pay off in later lessons.

ARROW 3: PACING YOUR PRESENTATION AND INTERACTING WITH STUDENTS AS THEY WORK

You should pace your verbal presentation at a normal speaking rate—as if you were telling somebody something important.

The arrows for number 3 on page 16 show two ways to pace your presentation for activities. One is marked with a ✔. The other is a note to **(Observe students and give feedback.)**. Both indicate that you will interact with students.

A ✔ is a note to check student performance on a task that requires only a second or two. If you are positioned close to several lower-performing students, quickly check whether they are responding appropriately. If they are, proceed with the presentation.

The **(Observe students and give feedback.)** direction requires more careful observation. You sample more students and you give feedback, not only to individual students but to the group. Here are the basic rules for what to do and what not to do when you observe and give feedback.

- If the task is one that takes no more than 30 seconds, observe and give feedback to several lower performing students.

- If the task requires considerably more time, move from the front of the room to a place where you can quickly sample the performance of low, middle, and high performers.

- As you observe, make comments to the whole class. Focus these comments on students who are following directions, working quickly, and working accurately.

"Wow, a couple of students are almost finished. I haven't seen one mistake so far."

- Students put their pencils down to indicate that they are finished.

Acknowledge students who are finished. They are not to work ahead.

- If you observe mistakes, do *not* provide a great deal of individual help. Point out any mistake, but do not work the problems for the students. For instance, if a student gets one of the problems wrong, point to the problem and say, "You made a mistake." If students don't line up their numerals correctly, say, "You'd better erase that and try again. Your numerals are not lined up." If students are not following instructions that you give, tell them, "You didn't follow my directions. You have to listen carefully. I said, 'Just work problem A; then stop.'" Make sure that you check the lower performers and give them feedback. When you show them what they did wrong, keep your explanation simple. The more involved your explanations, the more likely they are to get confused.

- If you observe a serious problem that is not unique to the lowest performers, tell the class, "Stop. We have a problem." Point out the mistake. Repeat the part of the exercise that gives them information about what they are to do. Do not provide new teaching or new problems. Simply repeat the part of the exercise that gives students the information they need and reassign the work. "Work it the right way."

- Allow students a reasonable amount of time. Do not wait for the slowest students to complete the problems before presenting the workcheck during which students correct their work and fix any mistakes. You can usually use the middle performers as a gauge for what is reasonable. As you observe that they are completing their work, announce, "Okay, you have about 10 seconds more to finish up." At the end of that time, begin the workcheck.

If you follow the procedures for observing students and giving feedback, your students will work faster and more accurately. They will also become facile at following your directions.

- If you wait a long time period before presenting the workcheck, you punish those who worked quickly and accurately. Soon, they will learn that there is no payoff for doing well—no praise, no recognition—but instead a long wait while you give attention to those who are slow.

- If you don't make announcements about students who are doing well and working quickly, the class will not understand what's expected. Students will probably not improve as much.

- If you provide extensive individual help on Independent Work, you will actually reinforce students for not listening to your directions and for being dependent on your help. Furthermore, this dependency becomes contagious. It doesn't take other students in the class long to discover that they don't have to listen to your directions, that they can raise their hand and receive help that shows them how to do the assigned work.

These expectations are the opposite of the ones you want to induce. You want students to be self-reliant and to have reasons for learning and remembering what you say when you instruct them. The simplest reasons are that they will use what they have just been shown and that they will receive reinforcement for performing well.

If you follow the management rules outlined above, all students who are properly placed in the program should be able to complete assigned work within a reasonable period of time and have reasons to feel good about their ability to do math. That's what you want to happen. As students improve, you should tell them about it. "What's this? Everybody's finished with that problem already? That's impressive."

ARROW 4: BOARDWORK/BOARD DISPLAYS

In many exercises, you will display problems on the board or a screen.

The word **(Display:)** appears in the script when a display is to be shown to students. The words **(Add to show:)** appear in the script when something is to be added to an existing display.

The program has been designed so that you can

a. show all displays and additions or changes to displays by using the Board Displays that come with the Teacher Presentation Book, or

b. write most displays on the board and project selected displays with an overhead projector or a document camera. Some displays will be difficult to write on a board.

Using the Board Displays to Show All Displays

The Board Displays contain all the displays for every lesson. The displays are labeled consecutively for each lesson. Note that the display code shown for each display begins with a number that indicates the lesson. The next number indicates the exercise on that lesson. The letters at the end of the code indicate the order of the displays. The code for the displays shown on page 16 are **61:4A, 4B,** and **4E** and **4F.** The **61** indicates Lesson 61. The **4** indicates that it is Exercise 4. The **A** indicates that it is the first display in that exercise. The **B** identifies the next display in Exercise 4. The first display for Exercise 5 is labeled [61:5A]. **Note:** The identification code appearing on each display corresponds to the code shown in the Presentation Book.

The best way to use the Board Displays is to stand where the images are projected and use a remote device to direct the presentation. (If you are using an interactive white board, you can simply touch the screen.) Being close to the image allows you to point to details of the display as you signal.

Using a Computer and Projector

You can move from display to display in several ways.

- You can use a remote control. Pressing the forward arrow on the remote calls up the next display. Pressing the back arrow returns to the previous display.

- You can also move from screen to screen by touching the right or left arrow on your computer's keyboard.

- You can use a mouse (wired or wireless) and click on the Next and Back arrows displayed on the computer screen.

- For an interactive white board, you can touch the Next and Back arrows at the bottom of the board.

Signaling with the Electronic Board Displays

The Board Displays have a built-in feature that can be used for signaling. The cursor shown on the screen can be replaced by an orange hand icon that, in turn, can be used to signal. You can replace the standard white cursor by clicking on the icon of a hand at the bottom of the screen. The cursor will turn into an orange hand-shaped cursor. When you click to signal, the hand cursor will move to simulate a "touch" that cues the students' response. Practice using the optional hand cursor before using it during a lesson.

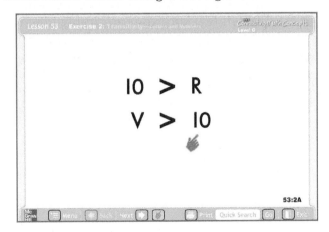

Other Methods for Presenting Displays

You can teach the program by writing most of the displays on the board. When the word **(Display:)** appears in the script, you simply write what appears in the script.

For example:

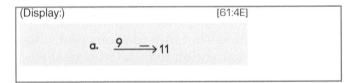

The Teacher Presentation Book shows changes to a display in white.

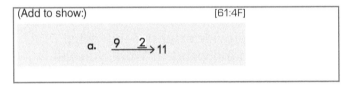

For complicated displays that would take time to write, you can print a copy of the display and use an overhead or document camera to project it.

Here's an example:

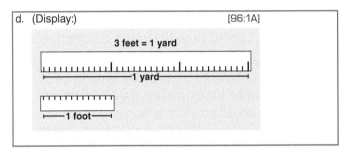

Preview lessons to determine if there are displays that would best be shown on a projector.

You can make printed copies of all the displays by accessing McGraw-Hill/SEG's ConnectED platform. Refer to the card you received with your teacher materials kit for more information on activating your six-year ConnectED subscription.

If you reproduce displays before the lesson, a document camera permits you to present them quickly during the lesson. This option is much more efficient than trying to write displays on the board.

Whatever system you use, your goal should be to keep the presentation moving without serious interruptions. If you are using the Board Displays, you will find that with practice you can present exercises at a good pace. If you are writing some displays on the board and projecting some displays, make sure you have the longer displays ready before presenting the lesson so you can maintain a good pace.

ARROW 5: WORKCHECKS (CHECK YOUR WORK.)

The purpose of the workcheck is to give students timely feedback on their work. It is important for students to correct mistakes on written work, and to do it in a way that allows you to see what mistakes they made.

Students should write their work in pencil so they can erase and make any corrections that are necessary as they work; however, they are not to change their work once it is finished.

Students are to use a colored pen or pencil for checking their work. When you indicate that it is time to "Check your work," they put down their pencils and pick up their marking pens. If their work is correct, they mark a **C** (or a $\checkmark$) for the item. If their work is wrong, they write the correct answer using their marking pens. Students hand in their marked-up lined-paper work and Workbooks daily. The errors that students make provide information on what needs to be firmed or repeated.

During workchecks that involve several problems, circulate among students and check their work. Praise students who are fixing mistakes. Allow a reasonable amount of time for them to check each problem.

Do not wait for the slowest students to finish their check. Keep the workcheck moving as quickly as possible.

INDEPENDENT WORK

The goal of the Independent Work is to provide review of previously taught content. The time required for Independent Work ranges from 10–25 minutes, with the time requirements increasing later in the program. Ideally, all Independent Work is completed in class, but not necessarily during the period in which the structured part of each lesson is presented. If it is not practical for students to complete the work at school, it may be assigned as homework. The cautions are that students are not to take their Textbook home. Students are not to tear out Workbook pages. They need to understand that if they take their Workbook home, they must bring it back the following school day.

Each newly introduced problem type becomes part of the Independent Work after it has appeared several times in structured teacher presentations. Everything that is taught in the program becomes part of the Independent Work. As a general rule, all major problem types that are taught in the program appear at least 10 times in the Independent Work. Some appear as many as 30 or more times. Early material is included in later lessons so that the Independent Work becomes relatively easy for students and provides them with evidence that they are successful.

Unacceptable Error Rates

Students' Independent Work should be monitored, and remedies should be provided for error rates that are too high. As a rule, if more than 30 percent of the students miss more than one or two items in any part of the Independent Work, provide a remedy for that part. The first lesson in which a recently taught skill is independent may have error rates of more than 30 percent of the students. Don't provide a remedy for these situations, but point out to the students that they had trouble with this part and possibly go over the most frequently missed problem. If an excessive error rate continues, provide a systematic correction.

High error rates on independent practice may be the result of the following:

a. The students may not be placed appropriately in the program.

b. The initial presentation may not have been adequately firmed. (The students made mistakes that were not corrected. The parts of the teacher presentation in which errors occurred were not repeated until firm.)

c. Students may have received inappropriate help. (When they worked structured problems earlier, they received too much help and became dependent on the help.)

d. Students may not have been required to follow directions carefully.

The simplest remedy for unacceptably high error rates on Independent Work is to repeat the structured exercises that occurred immediately before the material became independent. For example, if students have an unacceptable error rate on a particular kind of word problem, go to the last one or two exercises that presented the problem type as a teacher-directed activity. Repeat those exercises until students achieve a high level of mastery. Follow the script closely. Make sure you are not providing a great deal of additional prompting. Then assign the Independent Work for which their error rate was too high. Check to make sure students do not make too many errors.

GRADING PAPERS AND FEEDBACK

The teacher material includes a separate Answer Key. The key shows the work for all problems presented during the lesson and as Independent Work. When students are taught a particular method for working problems, they should follow the steps specified in the key. You should make sure students know that the work for a problem is wrong if the procedure is not followed.

After completing each lesson and before presenting the next lesson, follow these steps:

1. Check for excessive error rates for any parts of the written work from the structured part of the lesson. Note parts that have excessive error rates for more than 30% of the students. For instance, if a particular skill has had high error rates for more than two consecutive days, provide a remedy. Reteach as described above.

2. Conduct a workcheck for the Independent Work. One procedure is to provide a structured workcheck of Independent Work at the beginning of the period. Do not attempt to provide students with complete information about each problem. Read the answers. Students are to mark each item as correct or incorrect. The workcheck should not take more than five minutes. Students are to correct errors at a later time and hand in the corrected work. Keep records that show for each lesson whether students handed in corrected work. Attend to these aspects of the student's work:

    a. Were all the mistakes corrected?

    b. Is the appropriate work shown for each correction (not just the right answer)?

    c. Did the student perform acceptably on tasks that tended to be missed by other students? The answer to this question provides you with information on the student's performance on difficult tasks.

3. Award points for Independent Work performance. A good plan is to award one point for completing the Independent Work, one point for correcting all mistakes, and three points for making no more than four errors on the Independent Work. Students who do well can earn five points for each lesson. These points can be used as part of the basis for assigning grades. The Independent Work should be approximately one-third of the grade. The rest of the grade would be based on the Mastery Tests. (The Independent Work would provide students with up to 50 points for a ten-lesson period; each Mastery Test provides another possible 100 points—100% for a perfect test score.) An online progress monitoring application, *SRA 2Inform*, is available on ConnectED.

INDUCING APPROPRIATE LEARNING BEHAVIORS

Lower Performers

Here are other guidelines for reinforcing appropriate learning behaviors for lower performers who start at Lesson 1.

- During the first 10 lessons, hold students to a high criterion of performance. Remind them that they are to follow the procedures you show them.

- If they do poorly on the first Mastery Test, which follows Lesson 10, provide the specified Remedies (see the following section, In-Program Mastery Tests) then repeat Lessons 9 and 10. Tell students, "We're going to do these lessons again. This time, we'll do them perfectly."

Be positive. Reinforce students for following directions and not making the kinds of mistakes they had been making. Understand, however, that for some students, the relearning required to perform well is substantial, so be patient, but persistent. After students have completed lessons with a high level of success, they will understand your criteria for what they should do to perform acceptably. Retest them and point out that their performance shows that they are learning.

Following Directions

Students who have not gone through earlier levels of *CMC* may have strategies for approaching math that are not appropriate. Most notably, they may be very poor at following directions, even if they have an understanding of them. For instance, if you instruct the students to work problem C and then stop, a high percentage of them will not follow this direction.

Throughout *CMC Level C*, you will give students precise directions that may be quite different from those they have encountered earlier in their school experience. The most common problems occur when you direct students to work part of a problem and then stop, work one problem and then stop, or set up a series of problems without solving them.

The simplest remedy is to tell students early in the program that they are to listen to directions and follow them carefully. A good plan is to award points to the group (which means all members of the group earn the same number of points) for following directions. Praise students for attending to them. If you address the issue of following directions early in the school year, students will progress much faster later in the program, and you will not have to nag them as much about following directions.

Exercises that are the most difficult for these students, and the most difficult for the teacher to present effectively, are those that have long explanations. An effective procedure is to move fast enough to keep the students attentive.

Go fast on parts that do not require responses from the students; go more slowly and be more emphatic when presenting directions for what the students are to do. If you follow this general guideline, students will attend better to what you say.

Following Lined-Paper Icons

Students are to write answers to Textbook items on their lined paper. The Textbook is first used on Lesson 41. It is used in all subsequent lessons.

The general rules students follow are:

- They never write anything in their Textbook.
- They write their name at the top of the lined paper.
- They follow the layout icons shown in the Textbook.

Here are lined-paper icons from different Textbook parts. These icons show students how to lay out the problems on their lined paper.

For both parts, students write the letters a, b, and c for the items in the same form as they are shown in the Textbook.

For the first lined-paper icon, students write the item letters in a column.

For the second lined-paper icon, students write three letters on a row.

Expect students to require considerable practice to become proficient in reproducing the layouts for the different lined-paper icons. Specifically, some students won't skip lines according to the icon layout. Students often crowd items on a line close together; some students will write illegibly.

The work on following the layout shown by lined-paper icons is structured on Lessons 41 through 45.

Here's an example from Lesson 45:

a. (Hand out textbooks and lined paper.)
- Write your name at the top of your lined paper. ✔
b. Open your textbook to Lesson 45 and find part 1. ✔
(Teacher reference:)

You'll make a number family for each problem.
- Write **A** in front of the margin and make a number family arrow. ✔
- Count 4 lines and make the arrow for B. Then count lines and make the arrows for C and D. Your lined paper should look just like the picture in part 1.
(Observe students and give feedback.)

At this point, students should be practiced enough to follow these directions.

During the preceding lessons, hold students to a high criterion of performance. Do not permit them to set their paper up any way other than the scheme shown by the layout icon.

Following Solution Steps

An important behavior to address early is that students are to follow the solution steps that are taught. For much of their work, students make a number family, substitute numbers for one or more letters, write the number problem, and write the answer with a unit name. Students who follow these steps are able to solve problems far more reliably than students who do various calculations without properly representing the problem as a number family with two numbers.

Do not permit shortcuts or working the problem with steps missing. At first, this convention may strike some students as being laborious. Tell them early in the sequence that if they learn these steps, they will later avoid many difficulties students have when trying to work problems that involve a lot of steps. Point out to students that they will learn "real math strategies" that will permit them to have far less difficulty with higher math than they would if they were not well versed in solving problems by following the procedures you teach.

Long Lessons

Expect some lessons to run long. Do not hurry to complete a long lesson in one period if it would realistically take more than the period to present and check the material.

Complete a long lesson during the next period, and then start the next lesson during that period and go as far as you can during the allotted time. If you are running long on most of the lessons, the group may not be at mastery, or too much time is being spent on each problem set.

The simplest solution is to select one lesson and repeat it as many as two more times. If the students do not perform rapidly and accurately on all the exercises and work, they are misplaced in the program (which should have been apparent from Mastery Test performance). The best solution is to move the group back to an earlier lesson. Administer the previous two Mastery Tests to determine where students should be placed. Then reteach the program starting with the lesson for the new placement. Skip exercises that students have mastered.

In-Program Mastery Tests

Connecting Math Concepts Level C provides 13 in-program tests that permit you to assess how well each student is mastering the program content. The tests are packaged in the separate Student Assessment Book. Tests are scheduled to follow every tenth lesson, starting with Lesson 10 and continuing through the end of the program. The primary purpose of the tests is to provide you with information about how well each student is performing on the most recent things that have been taught in the program.

If the information shows that the group did not pass parts of the test, the program provides a specific remedy for each part. The Answer Key shows a passing criterion for each part of the test. A Remedy Table indicates the exercises that are to be repeated before the students are retested. Before presenting the next lesson, provide Remedies for parts students fail.

A good method for minimizing the possibility of students copying from each other is to maximize the space between them when they take the test. Discrepancies in the test performance and daily performance of some students pinpoint which students may be copying.

Below is Mastery Test 7, which is scheduled after Lesson 70.

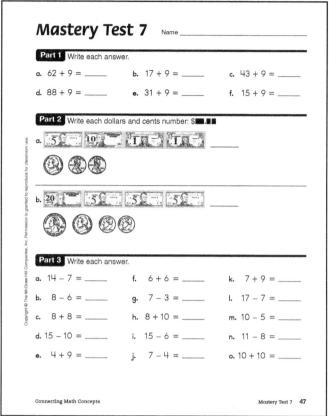

Mastery Test 7 Name _____

Part 1 Write each answer.

a. 62 + 9 = _____   b. 17 + 9 = _____   c. 43 + 9 = _____

d. 88 + 9 = _____   e. 31 + 9 = _____   f. 15 + 9 = _____

Part 2 Write each dollars and cents number: $■■.■■

a. _____

b. _____

Part 3 Write each answer.

a. 14 – 7 = _____   f. 6 + 6 = _____   k. 7 + 9 = _____

b. 8 – 6 = _____   g. 7 – 3 = _____   l. 17 – 7 = _____

c. 8 + 8 = _____   h. 8 + 10 = _____   m. 10 – 5 = _____

d. 15 – 10 = _____   i. 15 – 6 = _____   n. 11 – 8 = _____

e. 4 + 9 = _____   j. 7 – 4 = _____   o. 10 + 10 = _____

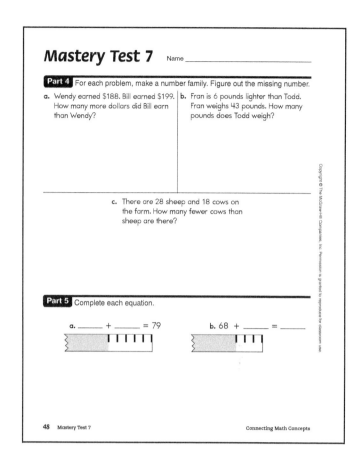

Mastery Test 7 Name _____

Part 4 For each problem, make a number family. Figure out the missing number.

a. Wendy earned $188. Bill earned $199. How many more dollars did Bill earn than Wendy?

b. Fran is 6 pounds lighter than Todd. Fran weighs 43 pounds. How many pounds does Todd weigh?

c. There are 28 sheep and 18 cows on the farm. How many fewer cows than sheep are there?

Part 5 Complete each equation.

a. _____ + _____ = 79

b. 68 + _____ = _____

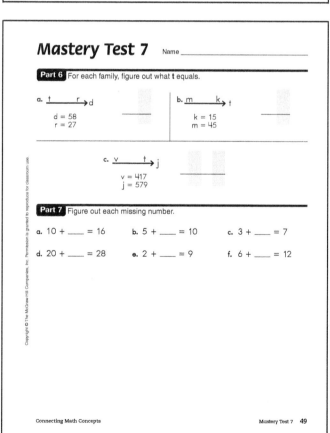

Mastery Test 7 Name _____

Part 6 For each family, figure out what **t** equals.

a. t ____r__→ d
    d = 58
    r = 27   _____

b. m ____k__→ t
    k = 15
    m = 45   _____

c. v ____t__→ j
    v = 417
    j = 579   _____

Part 7 Figure out each missing number.

a. 10 + ___ = 16   b. 5 + ___ = 10   c. 3 + ___ = 7

d. 20 + ___ = 28   e. 2 + ___ = 9   f. 6 + ___ = 12

Providing Remedies

The Answer Key for each Mastery Test provides the correct answers and shows the work for each item.

Tables that accompany each Mastery Test show the passing score for each part and indicate the percentages for different total test scores.

Here are the tables for Mastery Test 7:

	Passing Criteria Table — Mastery Test 7		
Part	Score	Possible Score	Passing Score
1	1 for each item	6	5
2	2 for each item (dollars, cents)	4	4
3	1 for each item	15	13
4	3 for each item (number family, problem, answer)	9	8
5	2 for each item	4	4
6	2 for each item (problem, answer)	6	5
7	1 for each item	6	5
	Total	50	

Test 7 Percent Summary					
Score	%	Score	%	Score	%
50	100	44	88	39	78
49	98	43	86	38	76
48	96	42	84	37	74
47	94	41	82	36	72
46	92	40	80	35	70
45	90				

The Passing Criteria Table gives the possible points for each item, the possible points for the part, and the passing criterion.

Students fail a part of the test if they score fewer than the specified number of passing points. For example, the total possible points for Part 1 is 6. A passing score is 5. If a student scores 5 or 6, the student passes the part. A student with a score of less than 5 fails the part.

Note that points are sometimes awarded for working different parts of the problem. For example, for Part 2, students earn 2 points for each item—one point for the dollar amount and one point for the cents amount.

The Percent Summary table shows the percentage grade you'd award students who have a perfect score of 50, a score of 49, and so forth.

Record each student's performance on the Group Summary of Mastery Test Performance (provided on pages 144–148 of this Teacher's Guide.) The Group Summary accommodates up to 30 students. The sample below shows only 6 students.

Here's how the results could be summarized following Mastery Test 7:

Remedy Summary—Group Summary of Test Performance									
Note: Test remedies are included in the Student Assessment Book. Percent Tables are provided in the Answer Key.				Test 7					
				Check parts not passed					Total %
Name	1	2	3	4	5	6	7		
1. Amanda		✔			✔		✔		
2. Karen			✔						
3. Adam	✔		✔			✔	✔		
4. Chan									
5. Felipe	✔								
6. Jack				✔			✔		
Number of students Not Passed = NP	2	0	4	0	1	1	3		
Total number of students = T	6	6	6	6	6	6	6		
Remedy needed if NP/T = 25% or more	Y	N	Y	N	N	N	Y		

The summary sheet provides you with a cumulative record of each student's performance on the in-program Mastery Tests.

Summarize each student's performance by making a check mark for each part failed.

At the bottom of each column, write the total number of failures for that part and the total number of students in the class. Then divide the number of failures by the number of students to determine the failure rate for each part.

Provide a group remedy for each part that has a failure rate of more than 25% (.25).

Test Remedies

The Answer Key specifies Remedies for each test. Any necessary Remedies should be presented before the next lesson (Lesson 71.)

Here are the remedies for Mastery Test 7:

Part	Test Items	Remedy Lesson	Remedy Exercise	Remedies Worksheet	Textbook
1	Mental Math (Plus 9)	66	2	Part A	—
		67	2	Part B	—
2	Money (Mixed Bills and Coins)	62	7	—	Part 3
		63	8	—	Part 3
3	Timed Facts* (Mixed Review)	62	1	Part C	—
		63	1	Part D	—
4	Comparison Word Problems (Mixed Set)	67	8	—	Part 2
		68	6	—	Part 2
5	Missing Addend (From Pictures)	69	1	Part E	—
		70	8 (step d)**	Part F	—
6	Number Families (With 3 Letters)	62	6	—	Part 2
		63	6	—	Part 1
7	Mental Math (Missing Addend)	63	3	Part G	—
		65	4	Part H	—

*Additional daily practice for the facts tested in part 3 are targeted in set 10 of the *Math Fact Worksheets.*
**Students work the problems immediately after the remedy exercise.

If the same students frequently fail parts of the Mastery Tests, it may be possible to provide Remedies for those students as the others do Independent Work. If individual students are weak on a particular skill, they will have trouble later in the program when that skill becomes a component in a larger operation or more complex application.

If students consistently fail tests, they are probably not placed appropriately in the program.

On the completed Group Summary of Test Performance for Mastery Test 7 on page 26, more than one quarter of the students failed Parts 1, 3, and 7. You provide group Remedies by re-teaching the exercises specified as the remedy for those parts. You may also need to provide individual Remedies for Amanda and Adam because they failed additional parts.

It may not be possible or practical to follow all the steps indicated for correcting individual students who make too many mistakes. For these cases, provide test Remedies to the *entire class* and move on in the program, attending to those students who make chronic mistakes without significantly slowing the group's progress.

Remedy Worksheets

All the Workbook parts needed for remedies appear in the Student Assessment Book immediately after each test. These are the Remedy worksheets. Below are remedy Parts A through D for Test 7, Parts 1 and 3.

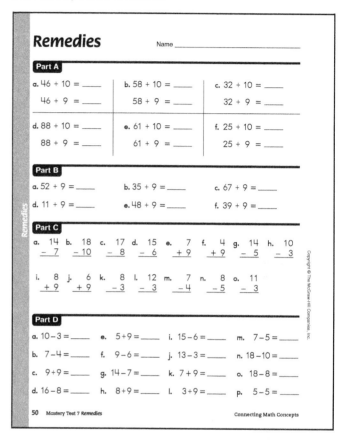

For Textbook remedies, students use their Textbook—not the Student Assessment Book. There are no Textbook remedies in the Student Assessment Book. Note that the remedy table specifies a Textbook remedy for Part 2 (Lesson 62, Textbook Part 3 and Lesson 63, Textbook Part 3).

Exercises that are specified as remedies for a mastery test have a REMEDY icon next to the exercise heading.

Remedy worksheet parts are also identified with an icon, e.g., R Part B. See page 153 for an example of a remedy exercise in sample Lesson 102.

Cumulative Tests

Level C has two Cumulative Tests. Cumulative Test 1 appears after Mastery Test 7 in the Student Assessment Book. Cumulative Test 2 appears after Mastery Test 13, at the end of the program.

Note that you don't have to present the Cumulative Test immediately after students complete Mastery Test 7. You may present this test any time before students reach Lesson 75 in the program.

These tests sample critical skills and concepts that have been taught earlier. Cumulative Test 1 assesses content from the beginning of the program through Lesson 70.

Cumulative Test 2 samples critical items from the beginning of the level, but it is heavily weighted for the lesson range from 70 to the end of the program. The presentation scripts for the Cumulative Tests appear at the end of each Presentation Book.

These Cumulative Tests may require more than one period to administer, but try to minimize, as much as possible, interrupting the schedule for teaching lessons.

Test remedies are not specified; however, the test scoring chart shows corresponding Mastery Test parts that may be used to identify possible remedies.

The Answer Keys for Cumulative Tests 1 and 2 follow.

Cumulative Test 1

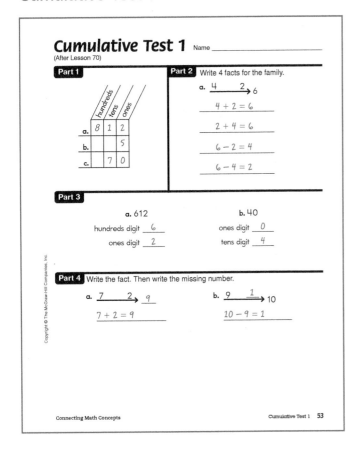

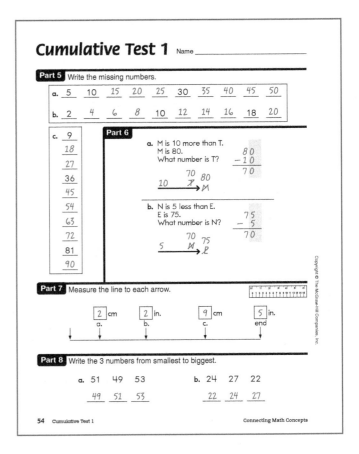

Cumulative Test 1 — Name _____

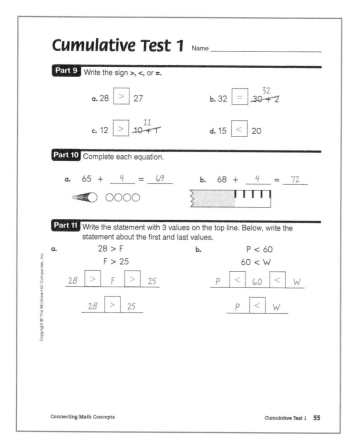

Part 9 Write the sign >, <, or =.

a. 28 [>] 27

b. 32 [=] 30 + 2 (32 written above)

c. 12 [>] 10 + 1 (11 written above)

d. 15 [<] 20

Part 10 Complete each equation.

a. 65 + _4_ = _69_

b. 68 + _4_ = _72_

Part 11 Write the statement with 3 values on the top line. Below, write the statement about the first and last values.

a.
28 > F
F > 25

28 [>] F [>] 25

28 [>] 25

b.
P < 60
60 < W

P [<] 60 [<] W

P [<] W

Cumulative Test 1 — Name _____

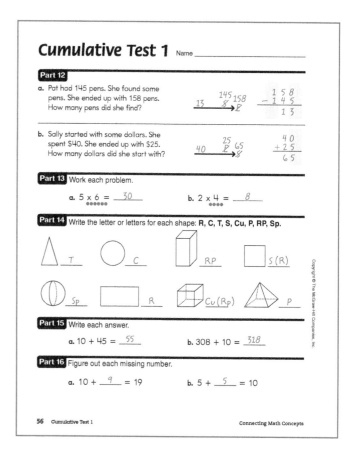

Part 12

a. Pat had 145 pens. She found some pens. She ended up with 158 pens. How many pens did she find?

13   145 → 158
      8 P

  1 5 8
− 1 4 5
  1 3

b. Sally started with some dollars. She spent $40. She ended up with $25. How many dollars did she start with?

40   25 → 65
      P 8

  4 0
+ 2 5
  6 5

Part 13 Work each problem.

a. 5 × 6 = _30_

b. 2 × 4 = _8_

Part 14 Write the letter or letters for each shape: R, C, T, S, Cu, P, RP, Sp.

_T_   _C_   _RP_   _S (R)_

_Sp_   _R_   _Cu (Rp)_   _P_

Part 15 Write each answer.

a. 10 + 45 = _55_

b. 308 + 10 = _318_

Part 16 Figure out each missing number.

a. 10 + _9_ = 19

b. 5 + _5_ = 10

Cumulative Test 1 — Name _____

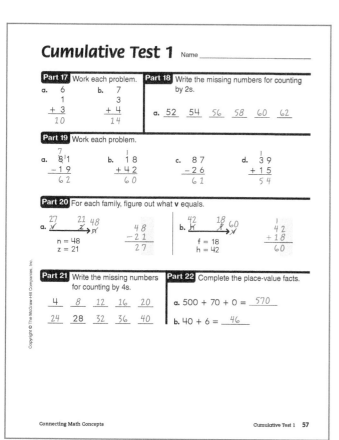

Part 17 Work each problem.

a.
  6
  1
+ 3
 10

b.
  7
  3
+ 4
 14

Part 18 Write the missing numbers for counting by 2s.

a. _52_ _54_ _56_ _58_ _60_ _62_

Part 19 Work each problem.

a.
  7
  8̸ 1
− 1 9
  6 2

b.
  1 8
+ 4 2
  6 0

c.
  8 7
− 2 6
  6 1

d.
  3 9
+ 1 5
  5 4

Part 20 For each family, figure out what **v** equals.

a.
27   21 → 48
 v   z n

n = 48
z = 21

  4 8
− 2 1
  2 7

b.
42   18 → 60
 h   f v

f = 18
h = 42

  4 2
+ 1 8
  6 0

Part 21 Write the missing numbers for counting by 4s.

_4_ _8_ _12_ _16_ _20_

_24_ _28_ _32_ _36_ _40_

Part 22 Complete the place-value facts.

a. 500 + 70 + 0 = _570_

b. 40 + 6 = _46_

Cumulative Test 2

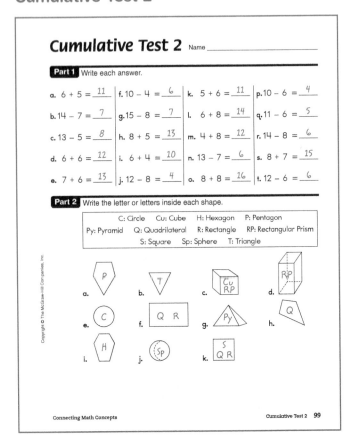

Cumulative Test 2 — Name _____

Part 1 Write each answer.

a. 6 + 5 = 11   f. 10 − 4 = 6   k. 5 + 6 = 11   p. 10 − 6 = 4

b. 14 − 7 = 7   g. 15 − 8 = 7   l. 6 + 8 = 14   q. 11 − 6 = 5

c. 13 − 5 = 8   h. 8 + 5 = 13   m. 4 + 8 = 12   r. 14 − 8 = 6

d. 6 + 6 = 12   i. 6 + 4 = 10   n. 13 − 7 = 6   s. 8 + 7 = 15

e. 7 + 6 = 13   j. 12 − 8 = 4   o. 8 + 8 = 16   t. 12 − 6 = 6

Part 2 Write the letter or letters inside each shape.

C: Circle   Cu: Cube   H: Hexagon   P: Pentagon
Py: Pyramid   Q: Quadrilateral   R: Rectangle   RP: Rectangular Prism
S: Square   Sp: Sphere   T: Triangle

a. P   b. T   c. Cu RP   d. RP

e. C   f. Q R   g. Py   h. Q

i. H   j. Sp   k. S Q R

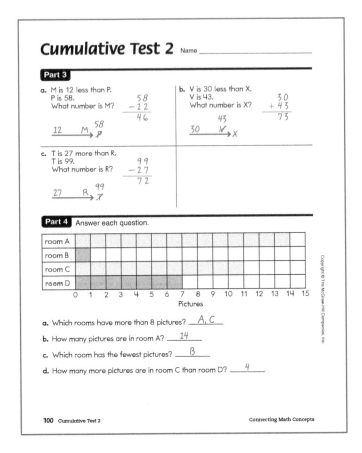

Cumulative Test 2 — Name _____

Part 3

a. M is 12 less than P.
P is 58.
What number is M?
58 − 12 = 46
12 → M → P 58

b. V is 30 less than X.
V is 43.
What number is X?
30 + 43 = 73
30 → V → X 43

c. T is 27 more than R.
T is 99.
What number is R?
99 − 27 = 72
27 → R → T 99

Part 4 Answer each question.

room A		
room B		
room C		
room D		

0 1 2 3 4 5 6 7 8 9 10 11 12 13 14 15
Pictures

a. Which rooms have more than 8 pictures? A, C

b. How many pictures are in room A? 14

c. Which room has the fewest pictures? B

d. How many more pictures are in room C than room D? 4

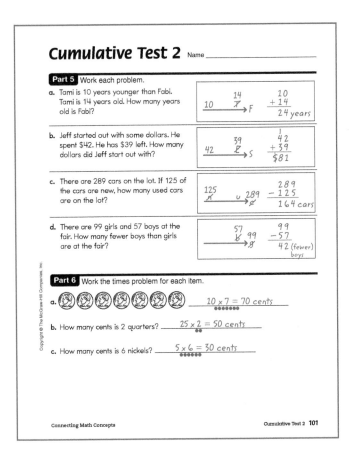

Cumulative Test 2 — Name _____

Part 5 Work each problem.

a. Tami is 10 years younger than Fabi. Tami is 14 years old. How many years old is Fabi?
10 → T → F   14   10 + 14 = 24 years

b. Jeff started out with some dollars. He spent $42. He has $39 left. How many dollars did Jeff start out with?
42 → R → S   39   42 + 39 = $81

c. There are 289 cars on the lot. If 125 of the cars are new, how many used cars are on the lot?
125 → U → 289   289 − 125 = 164 cars

d. There are 99 girls and 57 boys at the fair. How many fewer boys than girls are at the fair?
57 → B → 99   99 − 57 = 42 (fewer) boys

Part 6 Work the times problem for each item.

a. 10 × 7 = 70 cents

b. How many cents is 2 quarters? 25 × 2 = 50 cents

c. How many cents is 6 nickels? 5 × 6 = 30 cents

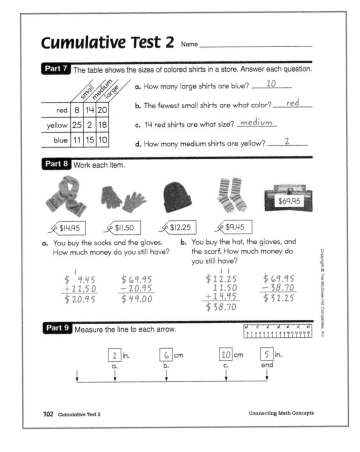

Cumulative Test 2 — Name _____

Part 7 The table shows the sizes of colored shirts in a store. Answer each question.

	small	medium	large
red	8	14	20
yellow	25	2	18
blue	11	15	10

a. How many large shirts are blue? 10

b. The fewest small shirts are what color? red

c. 14 red shirts are what size? medium

d. How many medium shirts are yellow? 2

Part 8 Work each item.

$14.95 $11.50 $12.25 $9.45 $69.95

a. You buy the socks and the gloves. How much money do you still have?
$9.45 + $11.50 = $20.95
$69.95 − $20.95 = $49.00

b. You buy the hat, the gloves, and the scarf. How much money do you still have?
$12.25 + $11.50 + $14.95 = $38.70
$69.95 − $38.70 = $31.25

Part 9 Measure the line to each arrow.

1 in.   6 cm   10 cm   5 in.
a.   b.   c.   end

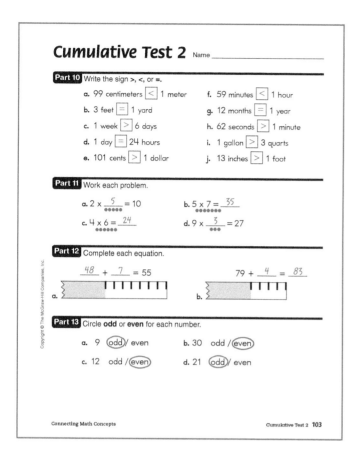

Cumulative Test 2 Name _____

Part 10 Write the sign >, <, or =.

a. 99 centimeters $\boxed{<}$ 1 meter
b. 3 feet $\boxed{=}$ 1 yard
c. 1 week $\boxed{>}$ 6 days
d. 1 day $\boxed{=}$ 24 hours
e. 101 cents $\boxed{>}$ 1 dollar
f. 59 minutes $\boxed{<}$ 1 hour
g. 12 months $\boxed{=}$ 1 year
h. 62 seconds $\boxed{>}$ 1 minute
i. 1 gallon $\boxed{>}$ 3 quarts
j. 13 inches $\boxed{>}$ 1 foot

Part 11 Work each problem.

a. 2 x $\underline{5}$ = 10
b. 5 x 7 = $\underline{35}$
c. 4 x 6 = $\underline{24}$
d. 9 x $\underline{3}$ = 27

Part 12 Complete each equation.

$\underline{48}$ + $\underline{7}$ = 55

79 + $\underline{4}$ = $\underline{83}$

a.
b.

Part 13 Circle **odd** or **even** for each number.

a. 9 (odd) / even
b. 30 odd / (even)
c. 12 odd / (even)
d. 21 (odd) / even

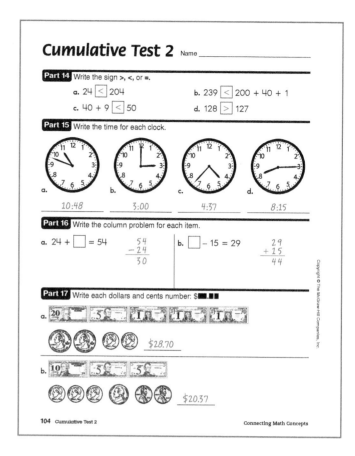

Cumulative Test 2 Name _____

Part 14 Write the sign >, <, or =.

a. 24 $\boxed{<}$ 204
b. 239 $\boxed{<}$ 200 + 40 + 1
c. 40 + 9 $\boxed{<}$ 50
d. 128 $\boxed{>}$ 127

Part 15 Write the time for each clock.

a. 10:48
b. 3:00
c. 4:37
d. 8:15

Part 16 Write the column problem for each item.

a. 24 + $\boxed{}$ = 54

$\begin{array}{r} 54 \\ -24 \\ \hline 30 \end{array}$

b. $\boxed{}$ − 15 = 29

$\begin{array}{r} 29 \\ +15 \\ \hline 44 \end{array}$

Part 17 Write each dollars and cents number: $■.■■

a. $28.70

b. $20.37

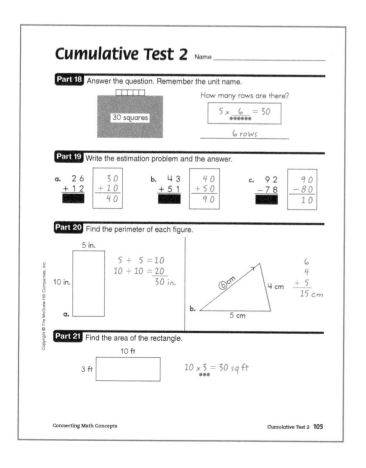

Cumulative Test 2 Name _____

Part 18 Answer the question. Remember the unit name.

30 squares

How many rows are there?

5 x $\underline{6}$ = 30

$\underline{6 \; rows}$

Part 19 Write the estimation problem and the answer.

a.
$\begin{array}{r} 26 \\ +12 \\ \hline \end{array}$
$\begin{array}{r} 30 \\ +10 \\ \hline 40 \end{array}$

b.
$\begin{array}{r} 43 \\ +51 \\ \hline \end{array}$
$\begin{array}{r} 40 \\ +50 \\ \hline 90 \end{array}$

c.
$\begin{array}{r} 92 \\ -78 \\ \hline \end{array}$
$\begin{array}{r} 90 \\ -80 \\ \hline 10 \end{array}$

Part 20 Find the perimeter of each figure.

5 in.

10 in.

a.

5 + 5 = 10
10 + 10 = 20
          30 in.

⑥cm

4 cm

b.

5 cm

$\begin{array}{r} 6 \\ 4 \\ +5 \\ \hline 15 \; cm \end{array}$

Part 21 Find the area of the rectangle.

10 ft

3 ft

10 x 3 = 30 sq ft

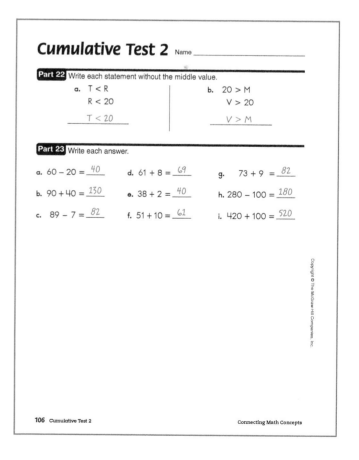

Cumulative Test 2 Name _____

Part 22 Write each statement without the middle value.

a. T < R
   R < 20
   $\underline{T < 20}$

b. 20 > M
   V > 20
   $\underline{V > M}$

Part 23 Write each answer.

a. 60 − 20 = $\underline{40}$
b. 90 + 40 = $\underline{130}$
c. 89 − 7 = $\underline{82}$
d. 61 + 8 = $\underline{69}$
e. 38 + 2 = $\underline{40}$
f. 51 + 10 = $\underline{61}$
g. 73 + 9 = $\underline{82}$
h. 280 − 100 = $\underline{180}$
i. 420 + 100 = $\underline{520}$

Tracks

Counting (Lessons 1–92)

The main emphasis of the counting activities is to prepare students to do mental calculations and to facilitate the teaching of a count-by strategy for multiplication (and multiplication facts in later levels).

The program assumes that entering students are able to count to 100 and to count backward from 10 to 1.

Counting forward and backward is reviewed in the first ten lessons, but not in a systematic manner.

The early counting activities focus on

- counting by tens and hundreds
- counting past 100 by ones
- counting backward from 20

The later counting sequences provide practice for counting by fives, twos, nines, and fours.

Students also learn to count by thousands and to count backward by tens, hundreds, and thousands.

On Lesson 1, students learn to count by tens. Here's the exercise:

a. Let's do some counting.
- Everybody, count from 1 to 10. (Signal.) *1, 2, 3, 4, 5, 6, 7, 8, 9, 10.*
b. (Display:)   [1:6A]

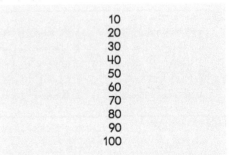

My turn to count by tens to 1 hundred.
- What am I going to count by? (Signal.) *Tens.*
Listen: 10, 20, 30, 40, 50, 60, 70, 80, 90, 1 hundred.
c. Listen to the first part: 10, 20, 30, 40, 50.
- Say that part with me. Get ready. (Signal.) *10, 20, 30, 40, 50.*
(Repeat until firm.)
d. Your turn: Count by tens to 50. Get ready. (Signal.) *10, 20, 30, 40, 50.*
- Listen: When you count by tens, what comes after 10? (Signal.) *20.*
- What comes after 20? (Signal.) *30.*
- What comes after 30? (Signal.) *40.*
- What comes after 40? (Signal.) *50.*
e. My turn to count by tens to 100: 10, 20, 30, 40, 50, 60, 70, 80, 90, 100.
- Count by tens to 100 with me. Get ready. (Signal.) *10, 20, 30, 40, 50, 60, 70, 80, 90, 100.*
(Repeat until firm.)

Lesson 1, Exercise 6

Teaching Note: In steps C and D, you model the counting, then "lead," and then "test."

When you lead, you count with the students.

When you test, the students count by themselves without your lead.

Make sure that you establish a rate for students to follow. It is very important for students to respond at this rate on the test step. If the group responds at a slower pace, the chances are good that one or two of the students are leading the group (initiating the responses) and the others are just following or mouthing the responses.

If you hold students to the same rate at which you modeled and led, all students will have to initiate the responses; in which case, you can see which students are having trouble.

If students are not firm, repeat the sequence or part that needs firming; however, don't repeat it more than three or four times.

If fewer than all the students are firm on the counting at the end of the exercise, there is no great problem because you will repeat these counting tasks on the following lessons. However, it's helpful for you to know which students are having trouble and exactly what that trouble is.

Here's the counting exercise from Lesson 3:

a. Listen: Count backward from 10. Get ready. (Signal.) *10, 9, 8, 7, 6, 5, 4, 3, 2, 1.*
b. My turn to count by tens to 1 hundred: 10, 20, 30, 40, 50, 60, 70, 80, 90, 1 hundred.
   • Your turn: Count by tens to 1 hundred. Get ready. (Signal.) *10, 20, 30, 40, 50, 60, 70, 80, 90, 1 hundred.*
   (Repeat until firm.)
c. When you count by tens, what comes after 40? (Signal.) *50.*
   • What comes after 70? (Signal.) *80.*
   • What comes after 30? (Signal.) *40.*
   • What comes after 50? (Signal.) *60.*
   (Repeat until firm.)
d. Once more: Count by tens to 1 hundred. (Signal.) *10, 20, 30, 40, 50, 60, 70, 80, 90, 1 hundred.*
e. My turn to count from 1 hundred to 1 hundred 10: 1 hundred, 1 hundred 1, 1 hundred 2, 1 hundred 3, 1 hundred 4, 1 hundred 5, 1 hundred 6, 1 hundred 7, 1 hundred 8, 1 hundred 9, 1 hundred 10.
   • Your turn: Count from 1 hundred to 1 hundred 10. Get ready. (Signal.) *100, 101, 102, 103, 104, 105, 106, 107, 108, 109, 110.*
   (Repeat until firm.)
f. Your turn: Count from 1 hundred 10 to 1 hundred 20. Get ready. (Signal.) *110, 111, 112, 113, 114, 115, 116, 117, 118, 119, 120.*
   (Repeat until firm.)

Lesson 3, Exercise 4

In step B, the teacher models counting by tens and then tests. (There is no lead step.)

In step D, students count without a model or a lead.

In step E, the teacher models and tests counting from 100 to 110. In step F, students count from 110 to 120.

Teaching Note: Lower performing students have trouble learning new patterns. Although counting from 100 to 110 is a simple verbal skill that requires saying "one hundred" before saying each of the familiar counting numbers (1–10), lower performers often require a lot of practice before they master any new pattern. It is important to bring them to mastery; but like the new counting tasks that begin on Lesson 1, mastery for all may not be a reasonable goal for the first day a new pattern is presented. Repeat the problem sequences three or four times, then move on, with the understanding that you'll work more on the problem sequences next time. Recognize, however, that students must master whatever patterns you introduce. If they do, they will be able to learn future patterns that are related with far less practice than the initial patterns required.

Note that students count backward from ten in step A. It's important for students to be firm on this counting because Lesson 5 introduces counting backward by tens from 100, which is closely related to the pattern for ones.

On Lesson 6, students start with 100 and count by hundreds to 1000. Here's part of the exercise that introduces this counting:

> d. My turn to count by hundreds to 1 thousand.
> - 1 hundred, 200, 300, 400, 500, 600, 700, 800, 900, 1 thousand.
> - What comes after nine hundred? (Signal.) *1 thousand.*
> e. I'll count and stop. You'll tell me the next number.
> - 600, 700, 800. What's the next number? (Signal.) *900.*
> - What's the next number? (Signal.) *1000.*
> f. Your turn: Start with 100 and count by hundreds to 1000. Get ready. (Signal.) *100, 200, 300, 400, 500, 600, 700, 800, 900, 1000.*
> (Repeat until firm.)

from Lesson 6, Exercise 3

Teaching Note: In the part of the exercise not shown, students review everything that has been presented since Lesson 1. This review pattern occurs throughout the program. Once something is taught, it is reviewed first daily and then intermittently. The program design makes the point to students that whatever is introduced will occur later in the program. Students should not have serious problems with counting by hundreds.

On Lesson 8, students are introduced to counting by fives. Here's part of the introduction:

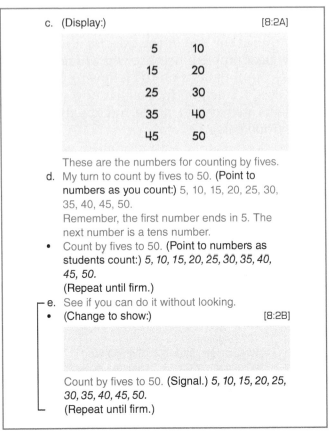

> c. (Display:)   [8:2A]
>
> | 5 | 10 |
> | 15 | 20 |
> | 25 | 30 |
> | 35 | 40 |
> | 45 | 50 |
>
> These are the numbers for counting by fives.
> d. My turn to count by fives to 50. (Point to numbers as you count:) 5, 10, 15, 20, 25, 30, 35, 40, 45, 50.
> Remember, the first number ends in 5. The next number is a tens number.
> - Count by fives to 50. (Point to numbers as students count:) *5, 10, 15, 20, 25, 30, 35, 40, 45, 50.*
> (Repeat until firm.)
> e. See if you can do it without looking.
> - (Change to show:)   [8:2B]
>
> Count by fives to 50. (Signal.) *5, 10, 15, 20, 25, 30, 35, 40, 45, 50.*
> (Repeat until firm.)

from Lesson 8, Exercise 2

Teaching Note: If students have trouble counting by fives to 50 in step D, model the counting with pauses after each pair of numbers:

5, 10 (pause), 15, 20 (pause)

and so forth. The pauses make the pattern more obvious, particularly with the pairs beyond 20.

Do not expect all the students to be able to count by fives to 50 when the display is removed (step E).

As soon as they falter, stop them. Then model just the first part, 5 through 30, and see if they are able to repeat that segment. If they can't after 3 or 4 trials, tell them to listen to the sequence again. Say the sequence; then tell students they'll work on it again next time. Remember, it's important for the students to hear the correct pattern as the last activity of this exercise.

On Lesson 12, students are introduced to counting by twos. Here's the exercise:

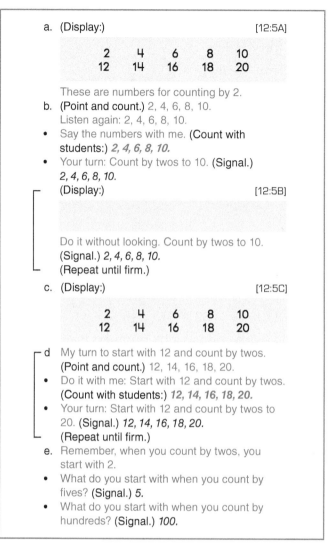

a. (Display:) [12:5A]

2	4	6	8	10
12	14	16	18	20

These are numbers for counting by 2.

b. (Point and count.) 2, 4, 6, 8, 10.
Listen again: 2, 4, 6, 8, 10.
• Say the numbers with me. (Count with students:) *2, 4, 6, 8, 10.*
• Your turn: Count by twos to 10. (Signal.) *2, 4, 6, 8, 10.*
(Display:) [12:5B]

Do it without looking. Count by twos to 10. (Signal.) *2, 4, 6, 8, 10.*
(Repeat until firm.)

c. (Display:) [12:5C]

2	4	6	8	10
12	14	16	18	20

d My turn to start with 12 and count by twos. (Point and count.) 12, 14, 16, 18, 20.
• Do it with me: Start with 12 and count by twos. (Count with students:) *12, 14, 16, 18, 20.*
• Your turn: Start with 12 and count by twos to 20. (Signal.) *12, 14, 16, 18, 20.*
(Repeat until firm.)

e. Remember, when you count by twos, you start with 2.
• What do you start with when you count by fives? (Signal.) *5.*
• What do you start with when you count by hundreds? (Signal.) *100.*

Lesson 12, Exercise 5

Here's a review exercise from Lesson 14.

a. My turn to count by twos to 20: 2, 4, 6, 8, 10 (pause) 12, 14, 16, 18, 20.
• Count by twos with me. (Count with students:) *2, 4, 6, 8, 10* (pause) *12, 14, 16, 18, 20.*
• Your turn: Count by twos to 20. (Signal.) *2, 4, 6, 8, 10, 12, 14, 16, 18, 20.*
(Repeat until firm.)

b. Count by fives to 50. Get ready. (Signal.) *5, 10, 15, 20, 25, 30, 35, 40, 45, 50.*
• Count by tens to 1 hundred. Get ready. (Signal.) *10, 20, 30, 40, 50, 60, 70, 80, 90, 100.*

c. What's 40 plus 10? (Signal.) *50.*
• What's 48 plus 10? (Signal.) *58.*
• What's 41 plus 10? (Signal.) *51.*
(Repeat until firm.)

d. My turn: What's 15 plus 10? 25.
• Your turn: What's 15 plus 10? (Signal.) *25.*
e. What's 17 plus 10? (Signal.) *27.*
• What's 13 plus 10? (Signal.) *23.*
• What's 11 plus 10? (Signal.) *21.*
(Repeat until firm.)

Lesson 14, Exercise 3

The teacher still models counting by twos. Students are expected to count by fives and tens without a model.

Students also review adding tens to different numbers.

Teaching Note: The numbers are arranged to show that there is a repeating pattern. The ones digits for this pattern are 2, 4, 6, 8, and 0.

When you present this counting, pause at the end of the row:

2, 4, 6, 8, 10 (pause) 12, 14, 16, 18, 20

Counting by nines begins on Lesson 15. Its pattern is that the tens digit increases by one and the ones digit decreases by one. Here's the first part of the exercise from Lesson 16:

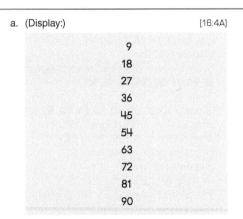

a. (Display:)   [16:4A]

9
18
27
36
45
54
63
72
81
90

These are the numbers you say when you count by nines to 90.
- Say them with me. (Point to each number as you and students count:) *9, 18, 27, 36, 45, 54, 63, 72, 81, 90.*

b. My turn to count by nines to 45: 9, 18, 27, 36, 45.
- Your turn: Count by nines to 45. (Signal.) *9, 18, 27, 36, 45.*
- (Repeat until firm.)

c. Remember, the ones digit of each number is 9, 8, 7, 6, 5.
- Again: Count by nines to 45. (Signal.) *9, 18, 27, 36, 45.*

d. (Display:)   [16:4B]

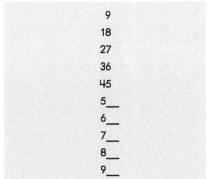

9
18
27
36
45
5__
6__
7__
8__
9__

- (Point to **45**.) What's the ones digit of 45? (Signal.) *5.*
  So the ones digit of the next number is 1 less than 5.
- What's 1 less than 5? (Signal.) *4.*

e. (Point to 5 __.) So what's the ones digit of this number? (Signal.) *4.*
- (Point to 6 __.) What's the ones digit of the next number? (Signal.) *3.*
- (Point to 7 __.) What's the ones digit of the next number? (Signal.) *2.*
- (Repeat until firm.)

f. My turn to start with 45 and count by nines to 90: 45, 54, 63, 72, 81, 90.
  Once more: 45, 54, 63, 72, 81, 90.
- Your turn: Start with 45 and count by nines to 90. (Signal.) *45, 54, 63, 72, 81, 90.*
- (Repeat until firm.)

from Lesson 16, Exercise 4

Teaching Note: Expect students to require considerable repetition of step F. The task is easier if you point to the digits that are shown and emphasize the ones digit in the numbers:

forty FIVE

fifty FOUR

sixty THREE

With this pattern, students are able to focus better on the pattern for the ones digits.

If they are still shaky after four repetitions, demonstrate the proper counting one time and remind them that they'll work on it more later.

The final counting series is 4s, which begins on Lesson 33. This pattern parallels the pattern for 2s:

4	8	12	16	20
24	28	32	36	40

The work with counting by numbers is reviewed intermittently throughout the level. It is also incorporated into multiplication problems. (See Multiplication track, page 87.)

COUNTING FROM A NUMBER

Students count from a number when they work coin problems, number line problems, and other types of count-on problems involving pictures.

Here's a picture showing two groups of coins.

Students count by tens for the dimes: 10, 20, 30. Then they continue counting by fives for the nickels: 35, 40, 45, 50.

Counting from a number other than one (such as 110–120) begins on Lesson 4. Skip-counting from a number begins on Lesson 23 after students are familiar with the basic skip-counting sequences (such as by fives to 50, by twos to 20). The students' knowledge of different counting patterns also permits the extension to numbers beyond those they originally practice; for instance, starting at 40 and counting by twos to 60 or starting with 55 and counting by fives to 70.

Here's part of an exercise from Lesson 78:

b. You'll start with 55 and count by fives to 95.
- What will you end with? (Signal.) *95.*
- Start with 55 and count by fives. (Signal.)
  55, 60, 65, 70, 75, 80, 85, 90, 95.
  (Repeat until firm.)
c. You'll start with 22 and count by twos to 40.
- What will you end with? (Signal.) *40.*
- Start with 22 and count by twos. (Signal.)
  22, 24, 26, 28, 30, 32, 34, 36, 38, 40.
  (Repeat until firm.)
d. You'll start with 36 and count by nines to 99.
- What will you end with? (Signal.) *99.*
- Start with 36 and count by nines. (Signal.)
  36, 45, 54, 63, 72, 81, 90, 99.
  (Repeat until firm.)
e. Listen: Start with 12 and count by fours to 48.
- What will you end with? (Signal.) *48.*
- Start with 12 and count by fours. (Signal.)
  12, 16, 20, 24, 28, 32, 36, 40, 44, 48.
  (Repeat until firm.)

from Lesson 78, Exercise 1

Place Value (Lessons 1–38)

Work on place value starts on Lesson 1. The early work continues through Lesson 24. Students later apply place-value analyses to read and write multi-digit numbers, to work column problems, and to draw inferences based on "expanded notation."

Students first learn to write numbers in columns that have column headings for hundreds, tens, and ones.

In the first part of the place-value exercise presented on Lesson 1, students learn what digits are and that different numbers have 1 digit, 2 digits, or 3 digits.

The exercise then presents column headings. Here's the part of the exercise that follows the reading of the column headings:

k. (Change to show:) [1:2G]

- (Point to **762**.) Everybody, read this number. (Touch.) *7 hundred 62.*
- How many digits are in 762? (Touch.) *Three.*
l. (Point to the word **hundreds**.) What's the hundreds digit in 7 hundred 62? (Touch.) *7.*
- (Point to the word **tens**.) What's the tens digit in 762? (Touch.) *6.*
- (Point to the word **ones**.) What's the ones digit in 762? (Touch.) *2.*
  (Repeat until firm.)

m. (Change to show:) [1:2H]

- (Point to **62**.) Read this number. (Touch.) *Sixty-two.*
- How many digits are in 62? (Touch.) *Two.*
n. (Point to the word **hundreds**.) Does 62 have a hundreds digit? (Touch.) *No.*
- (Point to the word **tens**.) Does 62 have a tens digit? (Touch.) *Yes.*
  What's the tens digit in 62? (Signal.) *6.*
- (Point to the word **ones**.) Does 62 have a ones digit? (Touch.) *Yes.*
  What's the ones digit in 62? (Signal.) *2.*

from Lesson 1, Exercise 2

Teaching Note: Initial work with place value is difficult because it requires a "double take." Students work with a number, but they also discuss the number of digits in that number.

In the first part of the exercise, students are told that if the number has 3 parts, it has 3 digits.

In steps K and M, you ask, "How many digits are in ___?" If students don't answer, point to the digits and count them: one, two, three.

Note that when you ask students to name the hundreds digit, the tens digit, or the ones digit, you point to the column headings. This prompts students to attend to the designation—hundreds, tens, and ones—rather than the number being analyzed.

Give special attention to this task when you practice before teaching the lesson.

Students write numbers in columns, starting on Lesson 1. Here's part of the exercise from Lesson 3:

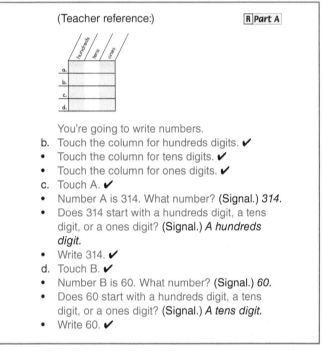

(Teacher reference:)   R Part A

You're going to write numbers.
b. Touch the column for hundreds digits. ✔
•   Touch the column for tens digits. ✔
•   Touch the column for ones digits. ✔
c. Touch A. ✔
•   Number A is 314. What number? (Signal.) *314.*
•   Does 314 start with a hundreds digit, a tens digit, or a ones digit? (Signal.) *A hundreds digit.*
•   Write 314. ✔
d. Touch B. ✔
•   Number B is 60. What number? (Signal.) *60.*
•   Does 60 start with a hundreds digit, a tens digit, or a ones digit? (Signal.) *A tens digit.*
•   Write 60. ✔

from Lesson 3, Exercise 8

The key question for each number is: Does ___ start with a hundreds digit, a tens digit, or a ones digit?

If students know the starting column and know how to write the number, they write it so it is properly aligned.

Teaching Note: If students make serious alignment mistakes, direct them to put an × next to each missed item before erasing it and rewriting the number properly.

After students have practiced writing regular 2-digit and 3-digit numbers, teen numbers are introduced. (This work begins on Lesson 13.) Monitor students' work carefully so you can make timely corrections. Some students may write these numbers backward—61 instead of 16. Other students, such as some native Spanish speakers, may write 60 instead of 16. This is because some native Spanish speakers tend to leave off the last consonant in some words, making it hard for them to discriminate between *sixteen* and *sixty.*

Expanded notation begins on Lesson 1. Students say the place-value facts for 2-digit numbers. For the first exercises, students read a number like 52 and say the "place-value fact:" 50 + 2 = 52. This work clarifies the relationship between the "parts" of the number and addition. (If you perform the addition 50 + 2, you get the answer 52.)

Another purpose of place-value facts is to clarify the relationship of tens digits and ones digits. If a number has two digits, the first digit has a tens value. This is true of numbers that some students confuse, particularly teen numbers. The addition clarifies the fact that the first digit has a tens value. This clarification helps students who tend to write teen numbers backward—41 for 14.

Here's part of the exercise from Lesson 3. Two pairs of examples are minimally different: 51 and 15; 18 and 81:

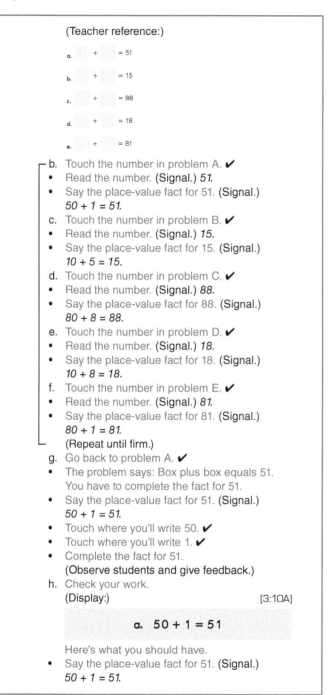

b. Touch the number in problem A. ✔
• Read the number. (Signal.) *51.*
• Say the place-value fact for 51. (Signal.)
  50 + 1 = 51.
c. Touch the number in problem B. ✔
• Read the number. (Signal.) *15.*
• Say the place-value fact for 15. (Signal.)
  10 + 5 = 15.
d. Touch the number in problem C. ✔
• Read the number. (Signal.) *88.*
• Say the place-value fact for 88. (Signal.)
  80 + 8 = 88.
e. Touch the number in problem D. ✔
• Read the number. (Signal.) *18.*
• Say the place-value fact for 18. (Signal.)
  10 + 8 = 18.
f. Touch the number in problem E. ✔
• Read the number. (Signal.) *81.*
• Say the place-value fact for 81. (Signal.)
  80 + 1 = 81.
  (Repeat until firm.)
g. Go back to problem A. ✔
• The problem says: Box plus box equals 51.
  You have to complete the fact for 51.
• Say the place-value fact for 51. (Signal.)
  50 + 1 = 51.
• Touch where you'll write 50. ✔
• Touch where you'll write 1. ✔
• Complete the fact for 51.
  (Observe students and give feedback.)
h. Check your work.
  (Display:)   [3:10A]

  a. 50 + 1 = 51

  Here's what you should have.
• Say the place-value fact for 51. (Signal.)
  50 + 1 = 51.

from Lesson 3, Exercise 10

Teaching Note: The exercise indicates that you repeat the verbal work (steps B through F) until students respond reliably on all the examples. If you make sure that students are firm, they probably will not make any serious mistakes when they complete the place-value facts. Note that you check these facts one at a time, starting at step H.

Expanded notation for 3-digit numbers begins on Lesson 13. For numbers like 342, students say the fact: 300 + 40 + 2 = 342. Here's the introduction:

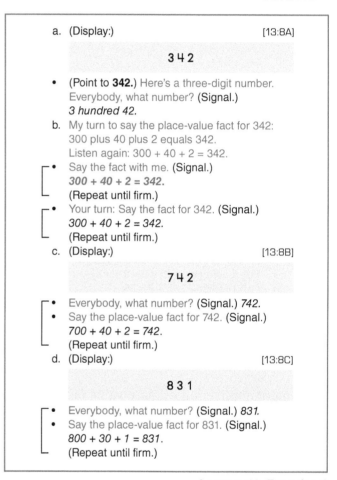

a. (Display:)   [13:8A]

  3 4 2

• (Point to **342**.) Here's a three-digit number. Everybody, what number? (Signal.)
  3 hundred 42.
b. My turn to say the place-value fact for 342:
  300 plus 40 plus 2 equals 342.
  Listen again: 300 + 40 + 2 = 342.
• Say the fact with me. (Signal.)
  300 + 40 + 2 = 342.
  (Repeat until firm.)
• Your turn: Say the fact for 342. (Signal.)
  300 + 40 + 2 = 342.
  (Repeat until firm.)
c. (Display:)   [13:8B]

  7 4 2

• Everybody, what number? (Signal.) *742.*
• Say the place-value fact for 742. (Signal.)
  700 + 40 + 2 = 742.
  (Repeat until firm.)
d. (Display:)   [13:8C]

  8 3 1

• Everybody, what number? (Signal.) *831.*
• Say the place-value fact for 831. (Signal.)
  800 + 30 + 1 = 831.
  (Repeat until firm.)

Lesson 13, Exercise 8

The final activity requires students to complete place-value facts. Students either write the addition or the answer. The teacher directions are minimal. Here's an activity from Lesson 24:

The major application of place-value facts occurs in connection with column subtraction. Students learn that for problems that require renaming, they rewrite the top number to show the "new" place-value fact. For example:

$$\begin{array}{r} 5 \\ \cancel{6}^{1}2 \\ -\ 1\ 3 \\ \hline \end{array}$$

Before working problems of this type, students learn to rewrite a standard place-value fact (e.g., 60 + 2 = 62 is rewritten as 50 + 12 = 62.) This provides a conceptual basis for rewriting a top number in the context of a subtraction problem. (See page 70.)

Number Families (Lessons 1–29)

Work on number families begins on Lesson 1 and continues throughout the program. Number families act as an organizational tool for fact memorization and for solving story problems.

Each family consists of an arrow and three numbers. The two numbers that are on the arrow are called small numbers. The number at the end of the arrow is called the big number. For most families, these three numbers generate four facts—two addition facts and two subtraction facts.

$$\underrightarrow{\quad 2 \qquad 3 \quad} 5 \qquad \underrightarrow{\quad 9 \qquad 4 \quad} 13$$

For the number family with small numbers of 2 and 3, the addition facts are 2 + 3 = 5 and 3 + 2 = 5. The subtraction facts are 5 – 3 = 2 and 5 – 2 = 3.

The main rationale for using number families to teach facts is that if students learn the arrangement of the three numbers in each family, they have a model for learning and remembering four facts. So the number families reduce the memory load required for students to become proficient with facts.

Another application of number families is solving number problems that are not facts. For this application, students apply two rules to number families:

If the big number is missing, you start with a small number and add to figure out the missing number:

$$\underrightarrow{\quad 23 \qquad 5 \quad} \_\_$$

This number family generates the addition problem 23 + 5.

If a small number is missing, you start with the big number and subtract to figure out the missing number:

$$\underrightarrow{\quad 23 \qquad \quad} 28$$

This number family generates the subtraction problem 28 – 23.

Students next learn to use letters in place of missing numbers, and work with problems that are not basic facts.

For the following number family, students work the problem 95 – 32 to figure out J:

$$\underset{\xrightarrow{\hspace{3cm}}}{\overset{32 \qquad \text{J}}{}} 95$$

For this number family, students work the problem 19 + 45 to figure out Z:

$$\underset{\xrightarrow{\hspace{3cm}}}{\overset{19 \qquad 45}{}} Z$$

Work with number problems prepares students for word problems. The only additional learning needed is to translate the word problems into a number family that has letters for the names and the numbers the problem gives. If the problem is translated properly, the family will permit students to write a number problem that solves the word problem. For example: Jane is 8 years older than Billy. Jane is 32 years old. How many years old is Billy?

The problem provides information about Jane and Billy. Jane is older, so J is the big number. Billy is younger, so he is one of the small numbers.

$$\underset{\xrightarrow{\hspace{3cm}}}{\overset{\text{B}}{}} J$$

The problem tells how much older Jane is—8 years. That number tells how much more or less. So it is a small number.

$$\underset{\xrightarrow{\hspace{3cm}}}{\overset{8 \qquad \text{B}}{}} J$$

The problem states that Jane is 32 years old. So we replace Jane with 32.

$$\underset{\xrightarrow{\hspace{3cm}}}{\overset{8 \qquad \text{B} \quad 32}{}} \cancel{J}$$

This family translates into the problem 32 – 8.

The application of number families to fact learning and word problems is discussed further in the Fact track (page 47) and the Word Problem track (page 53) that follow this section.

NUMBER FAMILY FOUNDATIONS

On Lesson 1, students learn that number families have three fact-related numbers. In the exercise, students say two addition facts—one that starts with the first small number and another that starts with the other small number.

Here's the exercise from Lesson 1:

a. (Display:) [1:3A]

$$\underset{\xrightarrow{\hspace{2cm}}}{\overset{6 \qquad 2}{}} 8$$
$$\underset{\xrightarrow{\hspace{2cm}}}{\overset{6 \qquad 3}{}} 9$$
$$\underset{\xrightarrow{\hspace{2cm}}}{\overset{6 \qquad 4}{}} 10$$

These are number families. Number families have three numbers that always go together. There are two small numbers and a big number. The small numbers are on the arrow. The big number is not on the arrow.

b. (Point to 6→2→8.) My turn: What are the small numbers in this family? 6 and 2.
• Your turn: What are the small numbers in this family? (Signal.) *6 and 2.*
c. My turn: What's the big number? 8.
• Your turn: What's the big number? (Signal.) *8.*
d. (Point to 6→3→9.) What are the small numbers in this family? (Signal.) *6 and 3.*
  What's the big number? (Signal.) *9.*
• (Point to 6→4→10.) What are the small numbers in this family? (Signal.) *6 and 4.*
  What's the big number? (Signal.) *10.*
  (Repeat until firm.)
e. You can say two addition facts for each family. One fact starts with the first small number.
• (Point to 6→2→8.) What's the first small number in this family? (Signal.) *6.*
f. Here's the fact: 6 plus 2 equals 8.
  Here's the fact that starts with the other small number: 2 plus 6 equals 8.
• Everybody, say the fact that starts with the first small number. (Signal.) *6 plus 2 equals 8.*
• Say the fact that starts with the other small number. (Signal.) *2 plus 6 equals 8.*
  (Repeat until firm.)
g. (Point to 6→3→9.) Everybody, say the fact that starts with the first small number. (Signal.) *6 plus 3 equals 9.*
• Say the other addition fact. (Signal.) *3 plus 6 equals 9.*
h. (Point to 6→4→10.) Everybody, say the fact that starts with the first small number. (Signal.) *6 plus 4 equals 10.*
• Say the other addition fact. (Signal.) *4 plus 6 equals 10.*
  (Repeat until firm.)
i. (Display:) [1:3B]

$$\underset{\xrightarrow{\hspace{2cm}}}{\overset{6 \qquad 1}{}} 7$$

• (Point to 6→1→7.) Everybody, say the fact that starts with the first small number. (Signal.) *6 plus 1 equals 7.*
• Say the other addition fact. (Signal.) *1 plus 6 equals 7.*
  (Repeat until firm.)

Lesson 1, Exercise 3

The examples are arranged so that they show incoming students the relationship between the three number families and familiar facts: 6 + 2 = 8, 2 + 6 = 8, 6 + 3 = 9, and so forth.

In steps B through E, students identify the two small numbers and the big number in each family.

In steps F through I, students say the two addition facts. The first fact begins with the first small number; the other fact begins with the second small number.

> **Teaching Note:** The brackets in the margin show exercise parts that are to be repeated if students make mistakes or don't respond well. Model exactly how you expect students to say the facts and tell them to say the facts just the way you do. This stipulation is important if the group is to become proficient at responding together.

On Lesson 2, students say subtraction facts for the families they worked on Lesson 1. Students learn the rule: All minus facts start with the big number.

Here's the part of the exercise that introduces new families:

m. (Display:)   [2:2B]

$$\underline{8 \quad \overset{2}{}}\!\!\rightarrow 10$$
$$\underline{9 \quad \overset{2}{}}\!\!\rightarrow 11$$
$$\underline{10 \quad \overset{2}{}}\!\!\rightarrow 12$$

Here are new number families.
n. You're going to say the minus facts. Remember, all minus facts start with the big number.
o. (Point to **8.**) What's the big number in this family? (Signal.) *10.*
• Say the minus fact that starts with 10 and goes backward along the arrow. (Signal.) *10 – 2 = 8.*
• Say the other minus fact. (Signal.) *10 – 8 = 2.*
p. (Point to **9.**) What's the big number in this family? (Signal.) *11.*
• Say the minus fact that goes backward along the arrow. (Signal.) *11 – 2 = 9.*
• Say the other minus fact. (Signal.) *11 – 9 = 2.*
q. (Point to **10.**) What's the big number in this family? (Signal.) *12.*
• Say the minus fact that goes backward along the arrow. (Signal.) *12 – 2 = 10.*
• Say the other minus fact. (Signal.) *12 – 10 = 2.*
(Repeat until firm.)
r. Remember, all addition facts start with a small number. All minus facts start with the big number.
• What do all addition facts start with? (Signal.) *A small number.*
• What do all minus facts start with? (Signal.) *The big number.*
(Repeat until firm.)

from Lesson 2, Exercise 2

> **Teaching Note:** Practice presenting this exercise before you work with students. Do the steps at a brisk pace. Remember to point before you talk.
>
> In step O, you point to 8 and ask, "What's the big number in this family?" By pointing to 8, you are not showing them the answer, just the number family that has a big number. Students have to look at the family and find the big number.
>
> If they make mistakes, repeat the step. Point:
>
> "What's the big number in this family? Say the minus fact that starts with 10 and goes backward along the arrow. Say the other minus fact."

Connecting Math Concepts

Lesson 4 introduces number families with missing numbers. Present the rule, "If the big number is missing, you add to figure it out."

Here's part of the exercise:

a. (Display:)   [4:6A]

$$\underline{5\quad}\to 8$$
$$\underline{5\quad 3}\to \_\_$$
$$\underline{=\quad 3}\to 8$$

d. Here's the rule: If the big number is missing, you add to figure it out.
• What do you do to find a missing big number? (Signal.) *Add.*
e. (Point to $\underline{5}\to 8$. Is the big number missing? (Signal.) *No.*
• So do you add to figure it out? (Signal.) *No.*
f. (Point to $\underline{5\quad 3}\to\_\_$.) Is the big number missing? (Signal.) *Yes.*
• So do you add to figure it out? (Signal.) *Yes.*
g. (Point to $\underline{=\quad 3}\to 8$.) Is the big number missing? (Signal.) *No.*
• So do you add to figure it out? (Signal.) *No.*
h. (Point to $\underline{5\quad 3}\to\_\_$.) Is the big number missing? (Signal.) *Yes.*
• So do you add? (Signal.) *Yes.*
• Start with 5 and say the problem. (Signal.) *5 + 3.*
(Repeat until firm.)

from Lesson 4, Exercise 6

On Lesson 5, students say subtraction problems for families that have a missing small number.

Here's part of the exercise:

a. (Display:)   [5:9A]

$$\underline{=\quad 1}\to 4$$
$$\underline{3\quad}\to 5$$
$$\underline{7\quad}\to 9$$
$$\underline{=\quad 8}\to 18$$

e. (Point to $\underline{=\quad 1}\to 4$.) Is the big number missing in this family? (Signal.) *No.*
• So do you add? (Signal.) *No.*
A small number is missing, so you subtract.
• What do you do to find a missing small number? (Signal.) *Subtract.*
• Here's the problem for this family: 4 – 1. Say the problem for this family. (Signal.) *4 – 1.*
f. (Point to $\underline{3\quad}\to 5$.) Is the big number missing in this family? (Signal.) *No.*
• Say the problem you work. (Signal.) *5 – 3.*
g. (Point to $\underline{7\quad}\to 9$.) Is the big number missing in this family? (Signal.) *No.*
• Say the problem you work. (Signal.) *9 – 7.*
h. (Point to $\underline{=\quad 8}\to 18$.) Is the big number missing in this family? (Signal.) *No.*
• Say the problem you work. (Signal.) *18 – 8.*
i. (Display:)   [5:9B]

$$\underline{4\quad}\to 9$$
$$\underline{10\quad 6}\to \_\_$$
$$\underline{=\quad 2}\to 10$$

Some of these families have a missing small number. Some have a missing big number.
j. Remember, if the big number is missing, you add.
• What do you do if the big number is missing? (Signal.) *Add.*
• What do you do if a small number is missing? (Signal.) *Subtract.*
k. (Point to $\underline{4\quad}\to 9$.) Is the big number missing in this family? (Signal.) *No.*
So do you add? (Signal.) *No.*
l. (Point to $\underline{4\quad}\to 9$.) Is the big number missing in this family? (Signal.) *No.*
• So do you add? (Signal.) *No.*
• Say the problem you work. (Signal.) *9 – 4.*
m. (Point to $\underline{10\quad 6}\to\_\_$.) Is the big number missing in this family? (Signal.) *Yes.*
• So do you add? (Signal.) *Yes.*

from Lesson 5, Exercise 9

Teaching Note: There are nested discriminations that students must learn if they are to become facile with number families. One is whether you add or subtract. If a small number is missing, you subtract. How do you construct that problem? You start with the big number, not a small number. Students require practice before they become adept at transforming number families into addition-subtraction problems.

Make sure that students are firm in steps E through H. They must say the problems without hesitation.

On Lesson 8, students identify whether the missing number is a small number or the big number. Then they say the problem, the answer, and the fact. (For problems with a missing number, students say two addition facts.)

Here's part of the exercise:

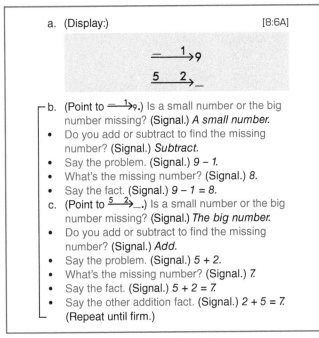

from Lesson 8, Exercise 6

Teaching Note: On the lesson, students carry out this routine with six more examples. This practice should be fast paced. This is the first time students say the facts, so make sure they respond correctly to the sequence of directions: "Say the problem," "What's the missing number?" and "Say the fact."

On the lessons through 14, students work with similar problems in their Workbook. Lesson 16 presents families that have two numbers and a letter. Students will work these problems the same way they work problems that have a missing number. They figure out the number for the letter, cross it out, and write the number answer.

Here's the second example presented on Lesson 16:

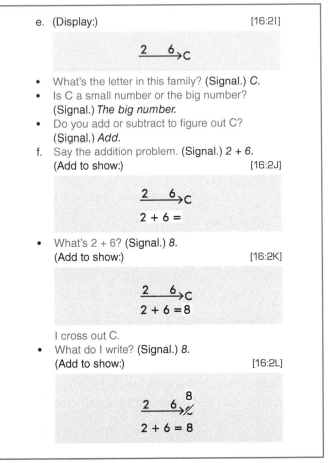

from Lesson 16, Exercise 2

You prompt the students with three questions:

- What's the letter in this family?
- Is C a small number or the big number?
- Do you add or subtract to figure out C?

Then you direct students to "Say the addition problem" and the answer.

You then cross out the letter and write the number above it.

This exercise presents the simplest context for students to work the problems. After they have practiced teacher-directed problems from displays, they work similar problems in their Workbooks.

Here is the exercise from Lesson 19:

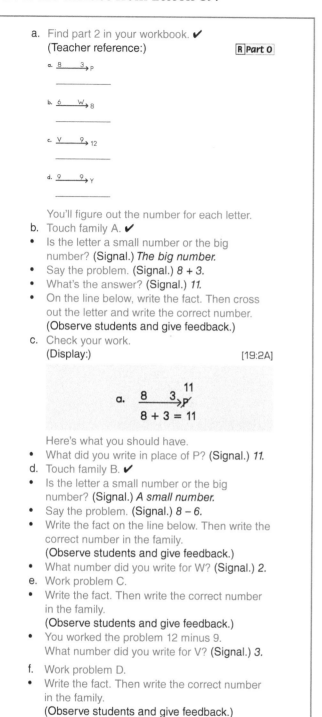

a. Find part 2 in your workbook. ✔
(Teacher reference:)

You'll figure out the number for each letter.
b. Touch family A. ✔
• Is the letter a small number or the big number? (Signal.) *The big number.*
• Say the problem. (Signal.) *8 + 3.*
• What's the answer? (Signal.) *11.*
• On the line below, write the fact. Then cross out the letter and write the correct number. (Observe students and give feedback.)
c. Check your work.
(Display:)   [19:2A]

Here's what you should have.
• What did you write in place of P? (Signal.) *11.*
d. Touch family B. ✔
• Is the letter a small number or the big number? (Signal.) *A small number.*
• Say the problem. (Signal.) *8 – 6.*
• Write the fact on the line below. Then write the correct number in the family. (Observe students and give feedback.)
• What number did you write for W? (Signal.) *2.*
e. Work problem C.
• Write the fact. Then write the correct number in the family. (Observe students and give feedback.)
• You worked the problem 12 minus 9. What number did you write for V? (Signal.) *3.*
f. Work problem D.
• Write the fact. Then write the correct number in the family. (Observe students and give feedback.)
• You worked the problem 9 plus 9. What number did you write for Y? (Signal.) *18.*

Lesson 19, Exercise 2

Students say the problem and the answer for family A. Then they write the problem and answer below the family, cross out the letter in the family, and write the number answer.

For the remaining problems, the teacher structures fewer steps. For C and D, the teacher simply directs students to "Write the fact. Then write the correct number in the family."

Starting on Lesson 22, students work problems that have two letters and a number. For example:

$$\underline{18 \qquad G}_{\,\searrow F}$$

Students can't work a number problem because there are two unknowns. The example gives information about one of the letters: G = 10. So G is crossed out in the number family and replaced by 10.

Here's that part of the exercise:

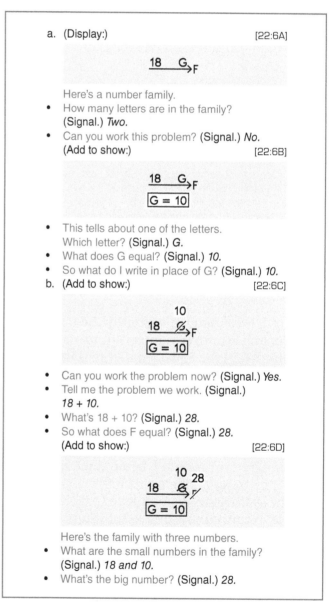

a. (Display:)   [22:6A]

Here's a number family.
• How many letters are in the family? (Signal.) *Two.*
• Can you work this problem? (Signal.) *No.*
(Add to show:)   [22:6B]

• This tells about one of the letters. Which letter? (Signal.) *G.*
• What does G equal? (Signal.) *10.*
• So what do I write in place of G? (Signal.) *10.*
b. (Add to show:)   [22:6C]

• Can you work the problem now? (Signal.) *Yes.*
• Tell me the problem we work. (Signal.) *18 + 10.*
• What's 18 + 10? (Signal.) *28.*
• So what does F equal? (Signal.) *28.*
(Add to show:)   [22:6D]

Here's the family with three numbers.
• What are the small numbers in the family? (Signal.) *18 and 10.*
• What's the big number? (Signal.) *28.*

from Lesson 22, Exercise 6

After two board examples, students work the same kind of problems as Workbook practice.

Here's the instruction for the first item:

a. Find part 3 in your workbook. ✔
(Teacher reference:)

a. $\underline{\quad 5 \quad C\quad}$→M   b. $\underline{\quad 20 \quad N\quad}$→K
   $\boxed{M = 7}$ $\boxed{N = 6}$

Each problem shows a number family with two letters.

b. Touch number family A. ✔
• What are the small numbers? (Signal.) *5 and C.*
• What's the big number? (Signal.) *M.*
• Touch the equation that tells about one of the letters. ✔
• Read the equation. (Signal.) *M = 7.*
• So what letter do you cross out in the number family? (Signal.) *M.*
• And what do you write above the crossed-out M? (Signal.) *7*
• Write the number for M in the number family. ✔

c. Check your work.
(Display:)   [22:6I]

$$\text{a.} \quad \underline{\quad 5 \quad C\quad} \overset{7}{\underset{\quad}{\longrightarrow}} M$$

Here's what you should have.
• Can you work this problem? (Signal.) *Yes.*
• Say the problem you work. (Signal.) *7 – 5.*
• What's the answer? (Signal.) *2.*
• Cross out C in the number family and write the number for C. ✔

from Lesson 22, Exercise 6

Teaching Note: Students identify the big number and the small numbers in the family. Then they read the equation that tells what M equals. They say what they will cross out in the family and what they will write above the crossed out letter. Then they fix up the family so it has two numbers and a letter.

It is important to make sure that students are very firm in this procedure because they will apply it to many problems in later lessons. Note that these problems are simple so that students are able to concentrate on the substitution procedure.

Later, students apply the procedures to problems that require more calculation.

Here are the problems from Lesson 27:

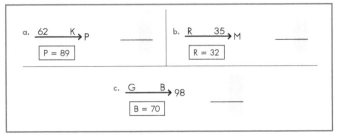

Workbook Lesson 27, Part 4

Here's the procedure specified for students working problem B:

e. Touch number family B. ✔
• The equation tells about a letter in the family. Cross out that letter and write the number in the family. ✔
(Display:)   [27:6D]

$$\text{b.} \quad \overset{32}{\underline{\quad R̶ \quad 35\quad}}→M$$
$$\boxed{R = 32}$$

Here's what you should have.

f. Write the problem you'll work to figure out the other letter in the family. Stop when you've written the problem. (**Observe students and give feedback.**)
(Add to show:)   [27:6E]

$$\text{b.} \quad \overset{32}{\underline{\quad R̶ \quad 35\quad}}→M \qquad \begin{array}{r} 3\ 2 \\ +\ 3\ 5 \\ \hline \end{array}$$
$$\boxed{R = 32}$$

Here's the problem you should have.
• Work the problem and put the number in the family. Raise your hand when you have a family with three numbers. ✔

g. Check your work.
(Add to show:)   [27:6F]

$$\text{b.} \quad \overset{32}{\underline{\quad R̶ \quad 35\quad}}\overset{67}{→M̶} \qquad \begin{array}{r} 3\ 2 \\ +\ 3\ 5 \\ \hline 6\ 7 \end{array}$$
$$\boxed{R = 32}$$

Here's what you should have.
• You figured out what M equals. What does M equal? (Signal.) *67.*

from Lesson 27, Exercise 6

Students first identify the letter that is to be replaced by a number. Next, they write the column problem for figuring out what M (the big number) equals. Finally, they cross out M and write what it equals. The result is a family with three numbers.

Students continue to work on column problems for number families with two letters through Lesson 29, at which time students work word problems that require the solution steps involving a number family with two letters. Solving these problems requires the steps shown above. Students make a number family, substitute a number for one of the letters, and then solve for the remaining letter.

Addition/Subtraction Facts (Lessons 1–115)

Work on facts begins on Lesson 1 and continues through Lesson 115. The first 64 lessons review facts that were taught in *CMC Level B*. These are facts from number families with small numbers of 1, 2, 3, 9, 10, and families for doubles ($4 \rightarrow 8$, $6 \rightarrow 12$)

Starting on Lesson 75, the remaining facts are introduced so that by the end of the program students will have worked with and practiced all the addition and subtraction facts through 20.

Groups of facts are taught in three stages. First, students work with a group of number families and practice saying the addition and subtraction facts for each family. Next, facts are grouped in written exercises in a way that prompts their connection to their number families. Finally, facts are presented in a random order that does not suggest their family ties. These final exercises are timed.

NUMBER FAMILIES FOR FACTS

Fact work involving number families begins on Lesson 9. After students have learned the basic properties of number families, the program introduces fact-learning exercises that are developed around number families.

On Lesson 9, students work with families that have a small number of 1.

Here's the first part of the exercise:

(Teacher reference:)

$\underset{1}{\rule{0.7cm}{0.4pt}}_{2} \quad \underset{2}{\rule{0.7cm}{0.4pt}}_{3} \quad \underset{3}{\rule{0.7cm}{0.4pt}}_{4} \quad \underset{4}{\rule{0.7cm}{0.4pt}}_{5} \quad \underset{5}{\rule{0.7cm}{0.4pt}}_{6} \quad \underset{6}{\rule{0.7cm}{0.4pt}}_{7} \quad \underset{7}{\rule{0.7cm}{0.4pt}}_{8} \quad \underset{8}{\rule{0.7cm}{0.4pt}}_{9} \quad \underset{9}{\rule{0.7cm}{0.4pt}}_{10} \quad \underset{10}{\rule{0.7cm}{0.4pt}}_{11} \quad \underset{12}{\rule{0.7cm}{0.4pt}}$

You're going to say facts for these families. All the families have 1 as a small number. So all the facts will have 1. Some will start with 1. Some will plus 1. Some will minus 1. And some will end with 1.

b. Touch the family with small numbers of 2 and 1. ✔
- Say the fact that starts with 1. (Signal.) *1 + 2 = 3.*
- Say the fact that goes backward along the arrow. (Signal.) *3 – 1 = 2.*
- Say the other minus fact. (Signal.) *3 – 2 = 1.* (Repeat until firm.)

c. Touch the next family. ✔
- Say the fact that starts with 1. (Signal.) *1 + 3 = 4.*
- Say the fact that goes backward along the arrow. (Signal.) *4 – 1 = 3.*
- Say the other minus fact. (Signal.) *4 – 3 = 1.*

d. Touch the next family. ✔
- Say the fact that starts with 1. (Signal.) *1 + 4 = 5.*
- Say the fact that goes backward along the arrow. (Signal.) *5 – 1 = 4.*
- Say the other minus fact. (Signal.) *5 – 4 = 1.* (Repeat until firm.)

from Lesson 9, Exercise 3

Later in the lesson, students work problems based on families that have a small number of 1.

Here's the set of problems students work and the teacher directions for working them:

(Teacher reference:)

a. $1 + 5 =$____	h. $5 + 1 =$____	o. $9 - 1 =$____
b. $10 - 9 =$____	i. $10 - 9 =$____	p. $1 + 3 =$____
c. $5 - 1 =$____	j. $3 - 1 =$____	q. $1 + 5 =$____
d. $1 + 8 =$____	k. $8 - 7 =$____	r. $6 - 5 =$____
e. $6 - 5 =$____	l. $6 + 1 =$____	s. $1 + 4 =$____
f. $9 - 1 =$____	m. $1 + 8 =$____	t. $3 - 1 =$____
g. $1 + 4 =$____	n. $10 + 1 =$____	

All these facts come from number families that have a small number of 1. You'll say the fact for each problem in the first column.

b. Problem A: $1 + 5$. Say the fact. (Signal.)
$1 + 5 = 6$.
• Problem B: $10 - 9$. Say the fact. (Signal.)
$10 - 9 = 1$.
• Problem C: $5 - 1$. Say the fact. (Signal.)
$5 - 1 = 4$.
• Problem D: $1 + 8$. Say the fact. (Signal.)
$1 + 8 = 9$.
• Problem E: $6 - 5$. Say the fact. (Signal.)
$6 - 5 = 1$.
• Problem F: $9 - 1$. Say the fact. (Signal.)
$9 - 1 = 8$.
• Problem G: $1 + 4$. Say the fact. (Signal.)
$1 + 4 = 5$.
(Repeat until firm.)
c. Write answers to all the problems in part 2. (Observe students and give feedback.)
d. Check your work. Read each fact.
• Fact A. (Signal.) $1 + 5 = 6$.

from Lesson 9, Exercise 6

Students entering *CMC Level C* probably know the plus-1 and minus-1 facts. The reason the program relates these facts to number families is that number families will be the basis for all the other facts that students learn.

Teaching Note: In step B, students say all the facts in the first column orally.

The procedure is for you to say the problem: $1 + 5$.

Students say the whole fact: $1 + 5 = 6$.

When you say the problem, you are actually providing students with "think time" for figuring out the answer, so maintain a brisk pace. If students have trouble with a particular fact, repeat that fact: "Here's the whole fact $9 - 1 = 8$. Say the fact."

If students miss any items in the first column, repeat the column.

Do not permit finger counting. Tell students that they have to remember the three numbers that go together in each family. If students count on their fingers, tell them to fold their hands during the oral part of the exercise.

The simplest procedure to discourage students from counting on their fingers or making lines when they are writing answers to the problems is to encourage them to go fast. Praise students who read the problem to themselves and then quickly write the answer. Frequently remind students, "Think of the three numbers in the number family." Although the results of this effort will not be apparent immediately, it will have a strong influence later when students learn new facts. Because one family tells about four related facts, there is a savings in teaching time and memory load.

Students continue to work on facts from families with a small number of 1 through Lesson 11.

On Lesson 12, students begin work with families that have a small number of 2. Students relate familiar subtraction facts to new facts shown by the families. The familiar facts minus 2. The new facts are the subtraction facts that equal 2.

a. (Display:)   [12:8A]

$$\underline{3 \quad 2}_{\to 5}$$

$$\underline{6 \quad 2}_{\to 8}$$

$$\underline{9 \quad 2}_{\to 11}$$

$$\underline{4 \quad 2}_{\to 6}$$

$$\underline{8 \quad 2}_{\to 10}$$

$$\underline{10 \quad 2}_{\to 12}$$

These are number families with a small number of 2. You'll say two subtraction facts for each family.

b. (Point to **5**.) (Touch numbers as you say:)
   5 – 2.
   • Say the fact for 5 – 2. (Signal.) *5 – 2 = 3.*
   • So what's 5 – 3? (Signal.) *2.*
   • Say the fact. (Signal.) *5 – 3 = 2.*

c. (Point to **8**.) Listen: 8 – 2.
   • Say the fact. (Signal.) *8 – 2 = 6.*
   • So what's 8 – 6? (Signal.) *2.*
   • Say the fact. (Signal.) *8 – 6 = 2.*

d. (Point to **11**.) Listen: 11 – 2.
   • Say the fact. (Signal.) *11 – 2 = 9.*
   • So what's 11 – 9? (Signal.) *2.*
   • Say the fact. (Signal.) *11 – 9 = 2.*

e. (Point to **6**.) Listen: 6 – 2.
   • Say the fact. (Signal.) *6 – 2 = 4.*
   • So what's 6 – 4? (Signal.) *2.*
   • Say the fact. (Signal.) *6 – 4 = 2.*

f. (Point to **10**.) *Listen: 10 – 2.*
   • Say the fact. (Signal.) *10 – 2 = 8.*
   • So what's 10 – 8? (Signal.) *2.*
   • Say the fact. (Signal.) *10 – 8 = 2.*

g. (Point to **12**.) Listen: 12 – 2.
   • Say the fact. (Signal.) *12 – 2 = 10.*
   • So what's 12 – 10? (Signal.) *2.*
   • Say the fact. (Signal.) *12 – 10 = 2.*

from Lesson 12, Exercise 8

The verbal exercise is followed by a Workbook exercise that shows families with a missing small number. Students write the facts for the missing number.

Students continue to work with families that have a small number of 2 on the following lessons. On Lesson 17, students say both addition and subtraction facts for these families.

Here's part of the verbal work from Lesson 17:

a. Find the number-family table on the inside of the front cover of your workbook. ✔
   • Touch the row that has 2 as the second small number. ✔
   (Teacher reference:)

   $\underline{2\ \ 2}_{\to 4}\ \ \underline{2\ \ 2}_{\to 5}\ \ \underline{4\ \ 2}_{\to 6}\ \ \underline{5\ \ 2}_{\to 7}\ \ \underline{6\ \ 2}_{\to 8}\ \ \underline{7\ \ 2}_{\to 9}\ \ \underline{8\ \ 2}_{\to 10}\ \ \underline{9\ \ 2}_{\to 11}\ \ \underline{10\ \ 2}_{\to 12}$

   You're going to say facts for this row. All the families have **2** as a small number. So all the facts will have 2. Some will **start** with 2, some will **plus** 2, some will **minus** 2, and some will **end** with 2.

b. Touch the family with the small numbers of 3 and 2. ✔
   • Say the fact that starts with 2. (Signal.) *2 + 3 = 5.*
   • Say the fact that goes backward along the arrow. (Signal.) *5 – 2 = 3.*
   • Say the other subtraction fact. (Signal.) *5 – 3 = 2.*

c. Touch the next number family. ✔
   • Say the fact that starts with 2. (Signal.) *2 + 4 = 6.*
   • Say the fact that goes backward along the arrow. (Signal.) *6 – 2 = 4.*
   • Say the other subtraction fact. (Signal.) *6 – 4 = 2.*

d. Touch the next number family. ✔
   • Say the fact that starts with 2. (Signal.) *2 + 5 = 7.*
   • Say the fact that goes backward along the arrow. (Signal.) *7 – 2 = 5.*
   • Say the other subtraction fact. (Signal.) *7 – 5 = 2.*

from Lesson 17, Exercise 2

Later in the lesson, students write answers to addition and subtraction facts for families with a small number of 2.

Here's the first part of the exercise:

(Teacher reference:)

a. $2 + 5 =$ _____	k. $10 - 8 =$ _____
b. $7 - 2 =$ _____	l. $7 - 2 =$ _____
c. $7 - 5 =$ _____	m. $7 - 5 =$ _____
d. $2 + 8 =$ _____	n. $2 + 5 =$ _____
e. $10 - 2 =$ _____	o. $10 - 2 =$ _____
f. $10 - 8 =$ _____	p. $6 - 4 =$ _____
g. $2 + 4 =$ _____	q. $2 + 4 =$ _____
h. $6 - 2 =$ _____	r. $6 + 2 =$ _____
i. $6 - 4 =$ _____	s. $6 - 2 =$ _____
j. $4 + 2 =$ _____	t. $2 + 8 =$ _____

You'll say the fact for each problem in the first column. All these facts have the number 2 in them.

b. Problem A: 2 plus 5.
   Say the fact. (Signal.) $2 + 5 = 7$.
• Problem D: 7 minus 2.
   Say the fact. (Signal.) $7 - 2 = 5$.
• Problem C: 7 minus 5.
   Say the fact. (Signal.) $7 - 5 = 2$.
   (Repeat until firm.).

from Lesson 17, Exercise 9

Students say the facts for the problems in the first column then write answers to all the facts.

Teaching Note: The facts in the first column are ordered by family. This ordering prompts the relationship of family to facts. The problems in the second column are more randomized. For these, students have to remember the fact (or family).

Do not present the second column until students respond correctly to all the items in the first column.

TIMED EXERCISES

Starting on Lesson 20, students work on timed exercises for families with a small number of 2.

Here's the set of Workbook problems students work on Lesson 24, the last day of exercises for facts with a small number of 2:

a. $11 - 2 =$ _____	i. $11 - 9 =$ _____	q. $2 + 7 =$ _____	x. $3 + 2 =$ _____
b. $6 + 2 =$ _____	j. $10 - 8 =$ _____	r. $10 - 2 =$ _____	y. $9 - 2 =$ _____
c. $5 - 3 =$ _____	k. $7 + 2 =$ _____	s. $2 + 9 =$ _____	z. $8 + 2 =$ _____
d. $10 + 2 =$ _____	l. $6 - 4 =$ _____	t. $9 - 2 =$ _____	A. $12 - 10 =$ _____
e. $8 - 6 =$ _____	m. $7 - 5 =$ _____	u. $2 + 8 =$ _____	B. $5 + 2 =$ _____
f. $7 - 2 =$ _____	n. $9 + 2 =$ _____	v. $5 - 2 =$ _____	C. $4 - 2 =$ _____
g. $9 - 7 =$ _____	o. $2 + 2 =$ _____	w. $8 - 2 =$ _____	D. $9 - 7 =$ _____
h. $4 + 2 =$ _____	p. $12 - 2 =$ _____		

Workbook Lesson 24, Part 5

For this timing, students work 30 problems.

Before the timing, students respond verbally to the problems in the first column. Then the teacher directs the timing.

Here's part of the exercise:

a. Find part 5 in your workbook. ✔
• Pencils down. ✔
  (Teacher reference:)
  These are the same 30 problems you worked last time. They are in a different order.
b. You're going to say facts for the problems in the first column. Then you'll write answers to all the facts.
• Read problem A. (Signal.) *11 – 2.*
  What's 11 – 2? (Signal.) *9.*
• Read problem B. (Signal.) *6 + 2.*
  What's 6 + 2? (Signal.) *8.*
• Read problem C. (Signal.) *5 – 3.*
  What's 5 – 3? (Signal.) *2.*
• Read problem D. (Signal.) *10 + 2.*
  What's 10 + 2? (Signal.) *12.*
• Read problem E. (Signal.) *8 – 6.*
  What's 8 – 6? (Signal.) *2.*
• Read problem F. (Signal.) *7 – 2.*
  What's 7 – 2? (Signal.) *5.*
• Read problem G. (Signal.) *9 – 7.*
  What's 9 – 7? (Signal.) *2.*
• Read problem H. (Signal.) *4 + 2.*
  What's 4 + 2? (Signal.) *6.*
c. Now you'll write answers to all the problems in all the columns. You'll have **2** minutes. You'll have to go pretty fast to do all of them in 2 minutes.
d. Put your pencil on problem A. ✔
• Get ready. Start.
  (Observe students and give feedback for good efforts.)
  (Tell students when they have only 30 seconds left.)
e. (At the end of 2 minutes say:) Stop. Pencils down. ✔
• Raise your hand if you wrote answers to all the problems. ✔
• Raise your hand if you wrote answers to almost all the problems. ✔
• Put an **X** next to any problem that doesn't have an answer. ✔

from Lesson 24, Exercise 7

Students check their work. They put an X next to each wrong answer.

Here's the part of the exercise that follows:

g. Count all the Xs you made and write that number in the box at the top of part 5. If you didn't miss any facts, write zero in the box. **(Observe students and give feedback.)**
h. Look at the number you wrote in the box. ✔
• Raise your hand if you wrote zero in the box. ✔
  (Praise students who wrote zero.)
• Raise your hand if you wrote 1, 2, or 3 in the box. ✔
  You did very well.
i. If you got more than 3 facts wrong, practice the facts you missed, and you'll do better on the next timing.

from Lesson 24, Exercise 7

Teaching Note: The timings are based on 4 seconds per item. In other words, if students read each item and write the answer in 4 seconds or less, they will complete all the problems within the specified time limit. This time restriction should be appropriate for most of the students. If their classroom is heterogeneous, expect possibly 25 percent of the students to not complete the problem sets.

Additional practice worksheets are available for each series of facts in the Math Fact Worksheets (blackline masters available via ConnectED). Use these with the understanding that you should keep moving through the lessons at a rate of a lesson a day, but you should do what you can to provide additional practice for students who need it.

The same sequence of verbal exercises, written exercises, and timed written exercises continues for each new set of related facts. The following chart gives the lesson sequence for each new set of families introduced after Lesson 24.

Lessons	Fact Families	Examples
25–28	Small number of 10	8 + 10, 10 + 8, 18 – 10, 18 – 8
29–43	Small number of 3	7 + 3, 3 + 7, 10 – 3, 10 – 7
44–54	Small number of 9	5 + 9, 9 + 5, 14 – 5
55–60	Doubles	6 + 6, 12 – 6; 8 + 8, 16 – 8
73–77	Minus-9 facts	12 – 9, 14 – 9, 17 – 9
75–80	Small numbers of 5 and 4 and 6 and 4	5 + 4, 4 + 5, 9 – 4, 9 – 5
81–89	Small numbers of 6 and 5; 7 and 4; 7 and 5	7 + 4, 4 + 7, 11 – 4, 11 – 7
93–101	Small numbers of 7 and 6 and 8 and 6	7 + 6, 6 + 7, 13 – 6, 13 – 7
102–110	Small numbers of 8 and 4 and 8 and 5	8 + 5, 5 + 8, 13 – 5, 13 – 8
111–115	Small numbers of 8 and 7	8 + 7, 7 + 8, 15 – 7, 15 – 8

After each group of facts goes through the complete cycle, the facts are considered to be known facts and appear in various problems and applications that students work. For instance, after Lesson 43, facts with a small number of 3 appear in column problems, story problems, and independent fact-practice exercises.

Word Problems (Lessons 12–87)

When students work word problems, they translate the written problem into a number family that has letters and numbers. When they create a family that has two numbers and a letter, they solve for the letter. The number for this letter is the answer to the question the word problem posed.

Three types of problems are taught in *CMC Level C:* comparison problems, start-end problems, and classification problems.

- Comparison problems compare two entities: **Fran is 6 inches shorter than Todd. Fran is 46 inches tall. How many inches tall is Todd?**

- Start-end problems tell about a change that occurs: **A truck starts out with 38 packages. Then the truck picks up 12 more packages. How many packages end up on the truck?**

- Classification problems involve a subordinate class and a higher-order class, such as hammers and tools, cars and vehicles, hamburgers and food, boys and students: **There were 56 students at the beach. 29 were girls. How many were boys?**

These three types cover basic word problems and the requirements of the Common Core State Standards for Mathematics.

The instruction for each type occurs over an extensive lesson range and provides many examples of each problem type. We illustrate the overall instructional strategies with the comparison problems.

COMPARISON PROBLEMS

The component skills and discriminations for this problem type begin on Lesson 11. Complete word problems are introduced on Lesson 56, and structured work continues through Lesson 70.

The skills that students need to work comparison word problems are systematically developed in *CMC Level C.*

Comparison Sentences with Two Numbers

First, students do simple comparisons with numbers. For **10 is more than 7,** students conclude that 10 is the big number, and 7 is a small number. Students also work problems that refer to *less:* **4 is less than 5.**

---

a. I'm going to tell you about two numbers. One is the big number in the number family. The other is a small number in the family.

b. (Display:)   [11:7A]

> 10 is more than 7.
>
> ———→

- (Point.) Here's a sentence: 10 is more than 7. Say the sentence. (Signal.) *10 is more than 7.*
- Which number is bigger? (Signal.) *10.*
So I write 10 as the big number and 7 as the small number right next to 10.
(Add to show:)   [11:7B]

> 10 is more than 7.
>
> ——7→10

c. (Display:)   [11:7C]

> 4 is less than 5.
>
> ———→

- (Point.) New sentence: 4 is less than 5. Say the sentence. (Signal.) *4 is less than 5.*
- Which is bigger? (Signal.) *5.*
So 5 is the big number in the family.
- What small number is in that family? (Signal.) *4.*
So I write 5 as the big number and 4 as the small number.
(Add to show:)   [11:7D]

> 4 is less than 5.
>
> ——4→5

d. (Display:)   [11:7E]

> 56 is less than 60.
>
> ———→

- New sentence: 56 is less than 60. Say the sentence. (Signal.) *56 is less than 60.*
- Which is the big number in the family? (Signal.) *60.*
- Which is a small number in the family? (Signal.) *56.*
So I write 60 as the big number and 56 as the small number.
(Add to show:)   [11:7F]

> 56 is less than 60.
>
> ——56→60

from Lesson 11, Exercise 7

Next, students work with sentences that identify the other small number: **6 is 1 less than 7.** Students first put 6 and 7 in the family then write 1 as the first small number.

$$\xrightarrow{\quad 1 \qquad 6 \quad} 7$$

Comparison Sentences with One Letter

Comparison sentences with a letter are first presented on Lesson 12. This example compares a number and letter: **B is more than 19.** Students have to attend to what the sentence says about the values because one is not obviously more than the other.

The value that is more is the big number; the other value is the small number that is next to the big number.

---

a. (Display:)   [12:6A]

$$\blacksquare \xrightarrow{\qquad\qquad}$$

You're going to tell me where to write a letter and a number in a number family. The letter takes the place of a number, so listen carefully to the sentences because they tell if the number is more or the letter is more.

b. (Add to show:)   [12:6B]

B is more than 19.

$$\blacksquare \xrightarrow{\qquad\qquad}$$

Here's a sentence that tells which is more.
• B is more than 19.
  Say the sentence. (Signal.) *B is more than 19.*
• Which is more, B or 19? (Signal.) *B.*
  So B is the big number in the family, and 19 is a small number.
c. Listen again: B is more than 19. Which is more? (Signal.) *B.*
• So which is the big number in the family? (Signal.) *B.*
• And which is a small number in the family? (Signal.) *19.*

---

from Lesson 12, Exercise 6

Following this exercise, students complete number families in their Workbook. Here's the first example:

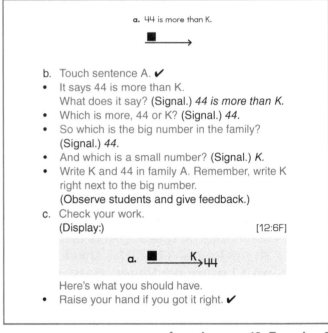

---

a. 44 is more than K.

$$\blacksquare \xrightarrow{\qquad\qquad}$$

b. Touch sentence A. ✔
• It says 44 is more than K.
  What does it say? (Signal.) *44 is more than K.*
• Which is more, 44 or K? (Signal.) *44.*
• So which is the big number in the family? (Signal.) *44.*
• And which is a small number? (Signal.) *K.*
• Write K and 44 in family A. Remember, write K right next to the big number.
  (Observe students and give feedback.)
c. Check your work.
  (Display:)   [12:6F]

a. $$\blacksquare \xrightarrow{\qquad\qquad K} 44$$

Here's what you should have.
• Raise your hand if you got it right. ✔

---

from Lesson 12, Exercise 6

Teaching Note: Students should have no trouble with this exercise. Attend to the mechanical details, particularly where students write K in the family. Praise students who write it very close to the arrow head.

Next, students work with combinations of letters and numbers.

First, one letter: **P is 3 more than 8.** P is more, so it is the big number. 8 and 3 are the small numbers.

$$\underline{\quad 3 \quad\quad 8 \quad}_{\searrow} P$$

Students use the same procedures for sentences like: **11 is 5 less than T.**

$$\underline{\quad 5 \quad\quad 11 \quad}_{\searrow} T$$

Starting on Lesson 16, students identify which number in a sentence tells how many more or less. For example: **45 is 32 more than P.** Students show which number tells more or less by circling 32.

On Lesson 21, the teacher makes number families with one letter and two numbers from sentences.

Here's part of the exercise:

```
c.  (Display:)                              [21:1G]

        35 is 20 more than J.

•  (Point to 35.) Read the sentence. (Signal.) 35
   is 20 more than J.
•  Which number tells how many more or less?
   (Signal.) 20.
   So I circle 20 and write it as the first small
   number in the family.
   (Add to show:)                           [21:1H]

        35 is⟨20⟩more than J.    ___20___→

•  Read the sentence without the circled number.
   (Signal.) 35 is more than J.
•  Which is the big number? (Signal.) 35.
•  Which is a small number? (Signal.) J.
   (Add to show:)                           [21:1I]

        35 is⟨20⟩more than J.  __20__ _J_→ 35

   Here's the family for the sentence: 35 is 20
   more than J.
```

from Lesson 21, Exercise 1

On Lesson 24, students work with sentences like: **23 is 41 less than F.** Students figure out what the letter equals.

Here's the part of the exercise that shows how the teacher directs students to perform the problem-solving steps:

```
f.  Sentence B: 23 is 41 less than F.
    Say the sentence. (Signal.) 23 is 41 less than F.
•   Circle the number that tells how many more or
    less. Then stop. ✔
•   Everybody, what number tells how many more
    or less? (Signal.) 41.
•   Write 41 as the first small number. ✔
g.  Read sentence B without the circled number.
    (Signal.) 23 is less than F.
•   Put 23 and F in the family. Then work a
    column problem to figure out what F equals.
    Remember to cross out F and write what it
    equals.
    (Observe students and give feedback.)
h.  Check your work.
    (Display:)                             [24:6D]

                        64          41
       b.  __41__ __23__→F̶       + 2 3
                                    ────
                                     6 4

    Here's what you should have.
•   What does F equal? (Signal.) 64.
```

from Lesson 24, Exercise 6

Students circle the number that tells *how many more* or *less*. They write that number as the first small number. Then they read the sentence without the circled number and identify F as the big number. The problem they work to figure out F is 41 + 23.

Teaching Note: These procedures buttress against various misrules that students develop about solving word problems of this type. The procedures contradict abortive strategies such as: If the problem has the word *less*, you subtract.

Later, the steps that the teacher directs in this exercise will be faded and students will have to initiate them. So make sure that you are following the procedures for solving this type of problem and that students are reliably performing the steps in the specified sequence.

Comparison Sentences with Two Letters

Next, students work with sentences that have two letters and a number: **C is 41 less than K. K is 68. What number is C?**

Students make the family, and then cross out K and write what it equals.

$$41 \quad \underset{\overset{\longrightarrow}{K}}{\cancel{C}} \,^{68}$$

Students solve for C.

On Lesson 28, you direct students to solve an entire problem. The steps in the solution combine what students have learned about comparison sentences with what they have learned about substituting a number for a letter.

Here's one of the problems students work:

> **F is 12 more than B.**
> **B is 77.**
> **What number is F?**

Following is part of the exercise that directs the solution steps:

---

a. F is 12 more than B.
B is 77.
What number is F?

_____

$\longrightarrow$

These are problems that have two letters. You're going to figure out the number for one of those letters.

b. Touch problem A. ✔
    Listen to the first sentence: F is 12 more than B.

•   Your turn: Circle the number that tells how many more or less and make the number family for that sentence. Pencils down when you've done that much. (Observe students and give feedback.)
    (Display:)   [28:6A]

$$\text{a.} \quad \underline{12} \quad \underset{\overset{\longrightarrow}{F}}{B}$$

Here's what you should have.

•   The next sentence in the problem says: B is 77.
    That means B equals 77. So you can cross out B and write 77. Do it. ✔
    (Add to show:)   [28:6B]

$$\text{a.} \quad \underline{12} \quad \underset{\overset{\longrightarrow}{F}}{\cancel{B}} \,^{77}$$

•   How many letters does the number family have now? (Signal.) *1.*
•   So you can figure out the number for F. Do it and write that number in the family.
    (Observe students and give feedback.)
c. Check your work.
•   Everybody, what number is F? (Signal.) *89.*
    (Add to show:)   [28:6C]

$$\text{a.} \quad \underline{12} \quad \underset{\overset{\longrightarrow}{\cancel{F}}}{\cancel{B}} \,^{77}\,^{89} \qquad \begin{array}{r} 1\,2 \\ +\,7\,7 \\ \hline 8\,9 \end{array}$$

Here's the family with three numbers.

from Lesson 28, Exercise 6

The first sentence of the problem tells students how to make the number family with a number and two letters. The second sentence gives a number for one of the letters. The students cross out the letter B and write what it equals.

Students work the problem for figuring out what F equals:

$$\begin{array}{r} 1\,2 \\ +\,7\,7 \\ \hline \end{array}$$

Students write the answer in the family.

> **Teaching Note:** All word-problem types have similar problem-solving steps: Students make a number family; put in the values the problem gives; write the number problem; and solve for the remaining number in the family.

With this framework, students have the mechanical skills needed to solve the word problems. What comes first is practice with sentences that have names, *not* letters: **Ann is 4 years younger than Jim.**

Students write a family with two letters that stand for the names and a number.

$$\underrightarrow{\quad 4 \qquad A \quad} J$$

Students work with a variety of these comparative sentences involving *heavier-lighter, fewer-more, taller-shorter, younger-older.*

Comparison Word Problems with Number Families

Finally, full word problems are introduced:

> **The dog was 21 pounds heavier than the cat. The dog weighed 32 pounds. How many pounds did the cat weigh?**

One sentence in the problem provides information about how to make a number family with two letters and a number. Students identify that sentence (The dog was 21 pounds heavier than the cat.), make a number family for that sentence, put in the other number the problem gives—32 for dog—and work the problem: 32 – 21. The answer is 11 pounds.

Students work problems of this type through the end of the program.

Starting on Lesson 47, students make number families from sentences of the type that will later appear in problems. Here's part of the exercise:

a. (Display:) [47:5A]

> a. **Fran is 4 years older than Jan.**
>
> b. **The gray dog is 3 years younger than the white dog.**
>
> c. **The boat is 1 year older than Tom.**

I'll read the sentences. You'll tell me who is older and who is younger.
b. Sentence A: Fran is 4 years older than Jan.
• Who is older? (Signal.) *Fran.*
• Who is younger? (Signal.) *Jan.*
c. Sentence B: The gray dog is 3 years younger than the white dog.
• Which is older? (Signal.) *The white dog.*
• Which is younger? (Signal.) *The gray dog.*
d. Sentence C: The boat is 1 year older than Tom.
• Which is older? (Signal.) *The boat.*
• Which is younger? (Signal.) *Tom.*
e. Listen: The one that is older is the big number. The one that is younger is a small number.
f. (Change to show:) [47:5B]

> a. <u>F</u>ran is 4 years older than <u>J</u>an.

(Point to **A.**) Listen again: Fran is 4 years older than Jan.
• Who is older? (Signal.) *Fran.*
• So who is the big number? (Signal.) *Fran.*
g. I write the first letter of the name for Fran. (Add to show:) [47:5C]

> a. <u>F</u>ran is 4 years older than <u>J</u>an.
>
> $\underrightarrow{\qquad\qquad} F$

• What's the letter for a small number? (Signal.) *J.* (Add to show:) [47:5D]

> a. <u>F</u>ran is 4 years older than <u>J</u>an.
>
> $\underrightarrow{\quad\quad J\quad} F$

• What number tells how many more or less? (Signal.) *4.* (Add to show:) [47:5E]

> a. <u>F</u>ran is 4 years older than <u>J</u>an.
>
> $\underrightarrow{\quad 4 \quad J\quad} F$

Here's the number family for sentence A: Fran is 4 years older than Jan.

from Lesson 47, Exercise 5

The family has two letters (one for each name) and a number. The rules for the arrangement of letters and number are the same as those for earlier problems with two letters and a number. In step F, The first letter of each name is underlined. This is a prompt for what students will write in the number family—**F** for Fran and **J** for Jan.

In the first steps of this exercise, the teacher reads each sentence and asks which of the things named in the sentence is older.

In step E, the teacher reminds students that the one that is older is the big number, and the one that is younger is the small number.

In steps F through G, the teacher directs the steps for making a number family for the sentence.

Teaching Note: The goal of the sentence exercises is not to solve a problem but to translate it into a number family with two letters. When students later work complete problems, they will identify the part of the word problem that provides a number for one of the letters. They then cross out the letter and write the number.

Complete Comparison Word Problems

On Lesson 48, students work complete problems. Here's part of the exercise:

(Teacher reference:)

a. Heidi has 17 more marbles than Bill has.
   Heidi has 48 marbles.
   How many marbles does Bill have?

b. Sarah made 10 more cupcakes than Maria made.
   Maria made 24 cupcakes.
   How many cupcakes did Sarah make?

c. Hank's car is 8 years older than Tim's car.
   Hank's car is 11 years old.
   How old is Tim's car?

d. Bob has 7 dollars less than Val has.
   Bob has 31 dollars.
   How many dollars does Val have?

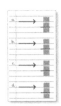

You're going to work problems that tell about people and things. Remember, the first letter of each name is underlined.

b. Touch problem A. ✔
• Heidi has 17 more marbles than Bill has. Say the sentence. (Signal.) *Heidi has 17 more marbles than Bill has.*
• What's the letter for Heidi? (Signal.) *H.*
• What's the letter for Bill? (Signal.) *B.*
  Heidi has 17 more marbles than Bill has.
• Make the family with two letters and the number. ✔
  (Display:)   [48:8A]

  a. ___17___ B→H

  Here's what you should have.

c. The next sentence in the problem says: Heidi has 48 marbles.
• Put a number for Heidi in the family. Then figure out how many marbles Bill has. Write that number in the family.
  (Observe students and give feedback.)
d. Check your work.
  (Add to show:)   [48:8B]

                   31 48   4 8
  a. ___17___ B̶→B̶1   – 1 7
                                    3 1

  Here's what you should have.
• The problem asks: How many marbles does Bill have? What's the answer? (Signal.) *31.*

from Lesson 48, Exercise 8

Teaching Note: At this point in the program, students have had a lot of practice with substituting numbers for letters and solving problems that have letters and a number. When working complete word problems, students must first configure the number family so it has two letters and a number. The key is that one of the sentences in the word problem tells how to construct the number family with two letters and a number. Once the family is constructed, students need to identify the number for one of the letters.

When correcting student mistakes, tell them about the big picture: "You make the family for one of the sentences. Then you find the number for one of the letters. You can't write a number problem until the family has two numbers and a letter."

On the following lessons, students make number families from sentences that have different adjectives: *heavier-lighter, taller-shorter, more-fewer.*

Here's part of a Workbook exercise from Lesson 50:

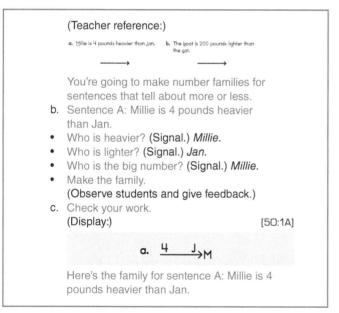

(Teacher reference:)

a. Millie is 4 pounds heavier than Jan.   b. The boat is 200 pounds lighter than the car.

You're going to make number families for sentences that tell about more or less.
b. Sentence A: Millie is 4 pounds heavier than Jan.
- Who is heavier? (Signal.) *Millie.*
- Who is lighter? (Signal.) *Jan.*
- Who is the big number? (Signal.) *Millie.*
- Make the family.
  (Observe students and give feedback.)
c. Check your work.
   (Display:)   [50:1A]

a. $\xrightarrow{\quad 4 \quad} \overset{J}{}{}_M$

Here's the family for sentence A: Millie is 4 pounds heavier than Jan.

from Lesson 50, Exercise 1

Students have learned that the entity that is heavier is the big number.

Teaching Note: Throughout the development of this skill, students should not have serious problems with any of the exercises. This does not mean that the material is "too easy" for the students, but simply that the students are receiving the kind of practice they need to become fluent with the components that will later occur in more complicated applications.

Complete Comparison Word Problems with Reduced Structure

On the following lessons, students work comparison word problems with reduced structure. Here's part of the exercise from Lesson 59:

(Teacher reference:)

a. A red barn is 18 feet taller than a white barn.
    The red barn is 38 feet tall.
    How many feet tall is the white barn?
b. There are 28 fewer green marbles than yellow marbles.
    There are 13 green marbles.
    How many yellow marbles are there?
c. The deck is 12 feet shorter than the fence.
    The fence is 36 feet long.
    How many feet long is the deck?
d. The stone wall is 8 feet longer than the brick wall.
    The brick wall is 23 feet long.
    How many feet long is the stone wall?

d. I'll read problem A: A red barn is 18 feet taller than a white barn. The red barn is 38 feet tall. How many feet tall is the white barn?
- Make the family and work the problem. Pencils down when you're finished.
  (Observe students and give feedback.)
e. Check your work.
   (Display:)   [59:7A]

a. $\xrightarrow[\quad 18 \quad]{\overset{20}{} \underset{38}{\cancel{M}}}$ $\begin{array}{r} 3\,8 \\ -1\,8 \\ \hline 2\,0 \end{array}$

Here's what you should have.
- The problem asks: How many feet tall is the white barn? Everybody, how many feet tall is the white barn? (Signal.) *20.*

from Lesson 59, Exercise 7

The problems involve *taller, fewer,* and *shorter.* The only structure the teacher provides is reading the problem and directing students to make the number family and work the problem.

Comparison Word Problems that Ask How Much More or Less

On Lesson 60, students begin work on the last type of comparison problem that involves sentences with numbers: **How much more is 12 than 2?**

Students make a number family with two numbers.

They work the problem for finding the first small number in the family: 12 − 2. The next sentence type starts on Lesson 63. This type has questions like: **How much shorter is Henry than Fran?** Students simply make the family with two letters. (They don't work a problem.)

Complete Comparison Word Problems that Ask How Much More or Less

On Lesson 65, complete problems are introduced. Here's part of the introduction:

(Teacher reference:)

a. Tom is 16 years old. Mary is 47 years old. How many years older is Mary than Tom?

b. Sam ran 18 miles. Fran ran 6 miles. How many more miles did Sam run than Fran?

c. Hillary is 67 inches tall. Ann is 50 inches tall. How many inches shorter is Ann than Hillary?

d. Vern had 6 blue shirts. Greg had 11 blue shirts. How many more blue shirts did Greg have than Vern had?

    Here's a new kind of word problem.
c. Touch problem A. ✔
    Tom is 16 years old. Mary is 47 years old. How many years older is Mary than Tom?
    Listen to the question again: How many years older is Mary than Tom?
•   Make a number family for that sentence. Stop when you've done that much.
    (Observe students and give feedback.)
    (Display:)   [65:7A]

    Here's what you should have.

d. The problem gives numbers for Mary and Tom. So cross out the letters and write the numbers above them. ✔
    (Add to show:)   [65:7B]

a. ——— 16 47 (crossed)

    Here's what you should have.
    You have two numbers, so now you can figure out how many years older Mary is.
•   Complete the number family.
    (Observe students and give feedback.)
e. Check your work.
    (Add to show:)   [65:7C]

a. 31 ——— 16 47 →M   47 − 16 = 31

•   Everybody, how many years older is Mary? (Signal.) 31.
f. Touch problem B. ✔
    Sam ran 18 miles. Fran ran 6 miles. How many more miles did Sam run than Fran?
•   Make the number family. Put in the two numbers the problem gives and figure out how many more miles Sam ran.
    (Observe students and give feedback.)
g. Check your work.
    (Display:)   [65:7D]

b. 12 ——— 6 18 →S   18 − 6 = 12

    Here's what you should have.
    Sam ran 12 miles more than Fran.

from Lesson 65, Exercise 7

Teaching Note: For this type of problem, the question provides the information students need to make the number family.

For the first problem, the teacher directs students to make the family for the question. For the remaining examples, the teacher does not refer to the question.

If students have trouble, tell them to read the question: **How many more miles did Sam run than Fran?** Then ask them,

"Does that question name both Sam and Fran?"

"Does that question tell you which is the big number?"

"So the question tells you how to make the number family."

"Make the family."

Starting on Lesson 67, students work problem sets in which some of the problems tell how many more or less and other problems ask how many more or less.

Here's the first part of the exercise:

(Teacher reference:)

a. Jane ran 11 miles farther than Ginger ran. Ginger ran 9 miles. How many miles did Jane run?

b. Jane rode a bike for 52 miles. Ginger rode for 36 miles. How many more miles did Jane ride than Ginger?

c. Bill is 11 inches taller than Fran. Bill is 73 inches tall. How many inches tall is Fran?

d. Al weighs 135 pounds. Janice weighs 110 pounds. How many pounds heavier is Al than Janice?

Some of these problems **ask** about how many more or less. Some problems **tell** about how many more or less.

b. Problem A: Jane ran 11 miles farther than Ginger ran. Ginger ran 9 miles. How many miles did Jane run?

- One of the sentences in that problem tells how to make the number family. Touch that sentence. ✔
  You should be touching the sentence Jane ran 11 miles farther than Ginger ran.

c. Problem B: Jane rode a bike for 52 miles. Ginger rode for 36 miles. How many more miles did Jane ride than Ginger?

- Touch the sentence that tells how to make the number family. ✔
- Read that sentence. (Call on a student. *How many more miles did Jane ride than Ginger?*)
  Yes, the sentence asks: How many more miles did Jane ride than Ginger?
  (Repeat until firm.)

---

from Lesson 67, Exercise 8

Teaching Note: If students have trouble, remind them that the sentence that tells them how to make the number family must have both names and must refer to *more* or *less*.

START-END WORD PROBLEMS

Start-End problems tell about a change that occurs with a single person or thing. The truck gets heavier. The person spends money. The water pours from the vessel.

The solution strategy that students follow is quite different from the procedures they use to solve comparative word problems. The key difference is that students don't write letters for anything named in the problem. Rather, they use the letters **S** (for start) and **E** (for end) to show the big number and a small number in the family. If the person or thing starts out with more than he or she ends with, **S** is the big number. If the person or thing ends with more, **E** is the big number and **S** is a small number.

Here's a basic problem: **Rob started out with 65 dollars. He spent 25 dollars. How many dollars did he end up with?**

Here are the solution steps:

1. Find the sentence that tells what happened to get more or less.

2. Make a number family with the letters **S** and **E** and a number the sentence provides.

3. Put in another number the sentence gives.

4. Solve for **S** or **E**.

For the problem involving Rob, the sentence that tells what happened for him to get more or less is: **He spent 25 dollars.**

If he spent money, he ended up with less than he started with. According to the sentence, he ended up with 25 dollars less. So the number family has E as a small number and S as the big number. 25 is the other small number.

$$\underline{25E}_{\rightarrow S}$$

The problem gives a number for start—65. That number replaces S.

Now the family generates the number problem: 65 – 25. The answer is 40. So $40 is the amount Rob ends with.

The same steps apply to the minimally different problem that generates an addition problem.

For example: **Rob started out with some money. He spent 25 dollars. He ended up with 80 dollars. How many dollars did he start out with?**

Using "key words" to solve this problem doesn't work. The start-end strategy, however, shows that the solution requires addition, not subtraction.

The sentence that tells what happened to get more or less is the same as the previous problem; however, the problem gives the end number, not the start number.

Here's the family with two numbers and a letter:

$$\underset{\underset{\longrightarrow S}{}}{\overset{80}{\underset{}{25 \quad \cancel{E}}}}$$

Students work the problem: 25 + 80. The answer is $105.

The basic strategy the program teaches is for students to figure out if the person or thing in the problem ends with more or ends with less. If students know where to position the letter for end in the number family, they can also position S. If the person or thing ends with more, E is the big number, and S is, therefore, a small number.

If the person or thing ends with less, E is a small number, and S is, therefore, the big number.

Here's part of an exercise that comes after students have learned that E can be either the big number or a small number. These examples are not complete problems, only sentences that tell what happened to get more or less.

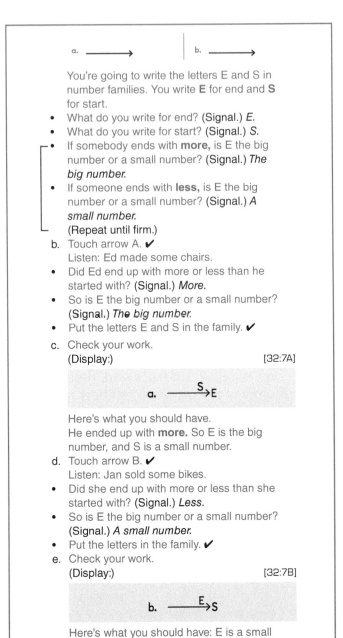

a. ⟶   b. ⟶

You're going to write the letters E and S in number families. You write **E** for end and **S** for start.
- What do you write for end? (Signal.) *E.*
- What do you write for start? (Signal.) *S.*
- If somebody ends with **more**, is E the big number or a small number? (Signal.) *The big number.*
- If someone ends with **less**, is E the big number or a small number? (Signal.) *A small number.*
  (Repeat until firm.)
b. Touch arrow A. ✔
   Listen: Ed made some chairs.
- Did Ed end up with more or less than he started with? (Signal.) *More.*
- So is E the big number or a small number? (Signal.) *The big number.*
- Put the letters E and S in the family. ✔
c. Check your work.
   (Display:)   [32:7A]

   a. ⟶ $\underset{\longrightarrow E}{S}$

Here's what you should have.
He ended up with **more**. So E is the big number, and S is a small number.
d. Touch arrow B. ✔
   Listen: Jan sold some bikes.
- Did she end up with more or less than she started with? (Signal.) *Less.*
- So is E the big number or a small number? (Signal.) *A small number.*
- Put the letters in the family. ✔
e. Check your work.
   (Display:)   [32:7B]

   b. ⟶ $\underset{\longrightarrow S}{E}$

Here's what you should have: E is a small number, and S is the big number.

from Lesson 32, Exercise 7

After students have worked with sentences that tell about start and end, they work with sentences that also provide a number for how many more or less (Tom lost 13 cards.)

On Lesson 36, students work complete problems. The procedure they follow is to underline the sentence that tells what happened to get more or less and make the number family based on that sentence.

Here's part of the exercise from Lesson 38.

---

c. Dolly had 8 bags.
She bought 30 more bags.
How many bags did she end up with?

f. Problem C. Follow along: Dolly had 8 bags. She bought 30 more bags. How many bags did she end up with?

• Underline the sentence that tells what happened to get more or less. ✔

⌐• Everybody, read the sentence you underlined. (Signal.) *She bought 30 more bags.* (Repeat until firm.)

g. Make a number family with a number and the letters E and S. (Observe students and give feedback.) (Display:)   [38:3E]

$$c. \quad \overset{30 \quad S}{\longrightarrow} E$$

Here's what you should have.

• Find the number of bags Dolly started with and put it in the family. ✔ (Add to show:)   [38:3F]

$$c. \quad \overset{8}{\underset{30 \quad \cancel{S}}{\longrightarrow}} E$$

Here's what you should have.

• Say the problem you'd work to figure out how many bags Dolly ended up with. (Signal.) *30 + 8.* Yes, 30 + 8.

• What's the answer? (Signal.) *38.*

h. Cross out E and write 38 in the number family. ✔ (Add to show:)   [38:3G]

$$c. \quad \overset{8 \quad 38}{\underset{30 \quad \cancel{S}}{\longrightarrow}\cancel{E}}$$

Here's what you should have.

from Lesson 38, Exericse 3

Teaching Note: Make sure that students underline the sentence that tells what happened to get more or less. Also make sure that students don't take shortcuts and work the problem without setting up the number family properly. Simple key-word strategies work for these problems. If the person gets less, it's a subtraction problem. If the person gets more, it's an addition problem. However, this formula does not work for problems that give the end number. Problems that give the end number are introduced on Lesson 39.

Here's part of the exercise from Lesson 40.

---

a. Bill started out with some marbles.
Then he gave 10 marbles away.
He ended up with 47 marbles.
How many marbles did he start out with?   ———

b. Touch problem A and follow along. ✔ Bill started out with some marbles. Then he gave 10 marbles away. He ended up with 47 marbles. How many marbles did he start out with?

• Underline the sentence that tells what happened to get more or less. ✔

• Everybody, read the sentence you underlined. (Signal.) *Then he gave 10 marbles away.*

• Make the family for that sentence. Then stop. (Observe students and give feedback.) (Display:)   [40:10D]

$$a. \quad \overset{10 \quad E}{\longrightarrow} S$$

Here's what you should have.

c. The problem gives a number for start or for end.

• Touch that number. ✔

• What number? (Signal.) *47.* Yes, he ended up with 47 marbles.

• Put the number for end in the family. ✔ (Display:)   [40:10E]

$$a. \quad \overset{47}{\underset{10 \quad \cancel{E}}{\longrightarrow}} S$$

• Everybody, say the problem you'll work. (Signal.) *10 + 47.*

• Write the problem. Then stop. ✔ You'll work the problem later.

from Lesson 40, Exercise 10

Students underline the sentence that tells what happened to get more or less and make a number family for that sentence.

In step C, students identify the number the problem gives for start or end. Students put that number in the family, and then say the problem they'll work to figure out the number for start.

Starting on Lesson 41, students work a mixed set of problems. Some give a number for start. Others give a number for end. Here's part of the exercise from Lesson 42 that presents minimally different problems. One gives a number for end. The other gives a number for start.

g. Problem C: A train had 97 people on it. Then 34 people got off the train. How many people ended up on the train?
• Make the number family with letters for start and end and with two numbers. ✔
(Display:)   [42:7C]

c. $\underset{\longrightarrow}{34}\ \underset{\mathscr{S}}{E}\overset{97}{}$

Here's what you should have.
h. Problem D: A train had some people on it. Then 155 people got off the train. The train ended up with 23 people. How many people did the train start with?
• Make the number family with letters for start and end and with two numbers. ✔
(Display:)   [42:7D]

d. $\underset{\longrightarrow}{155}\ \underset{\mathrm{S}}{\mathscr{E}}\overset{23}{}$

Here's what you should have.

from Lesson 42 Exercise 7

Teaching Note: After students set up the problems, they work them. The most important step to monitor is not the working of the problems, but how students set them up. If students do not follow the procedures precisely, they may have serious trouble later.

Make sure students are completely reliable in finding the sentence that tells what happened to get more or less, making the family for the sentence, and substituting a number for one of the letters in the family.

If students have trouble, repeat some of the problems presented on earlier lessons, and provide the structure students need to work these problems properly. Do not go farther in the program until students perform reliably.

Starting on Lesson 50, students work a new kind of start-end problem. It gives both the start number and the end number. Students figure out the number for the change that occurred.

For example: **A train started out with 46 people. Some more people got on the train. The train ended up with 58 people on it. How many got on the train?**

Students work the problem by making a number family with the two numbers the problem gives.

They work the problem 58 – 46.

The last mix of start-end problems begins on Lesson 51. Problems include those that ask about the start number, those that ask about the end number, and those that ask about how much more or less (the change). Here's the set of problems from Lesson 60:

a. Nan had some books. She gave away 93 books. She ended up with 78 books. How many books did Nan start with?

b. A store had 215 dolls. Then the store bought some more dolls. The store ended up with 368 dolls. How many dolls did the store buy?

c. A woman weighed 162 pounds. She lost some weight. She ended up weighing 120 pounds. How many pounds did the woman lose?

d. There were 15 dogs in the kennel. 19 more dogs came to stay. How many dogs ended up in the kennel?

from Textbook Lesson 60

Students work these problems with a minimum of teacher direction.

CLASSIFICATION WORD PROBLEMS

Starting on Lesson 63, students begin learning the component skills needed to solve the last type of word problem—classification problems.

Here's an example: **There are 56 people in the park. 21 of them are adults. How many are children?**

Before students work problems of this type, the program presents exercises on eight lessons that work on the classification skills. This work is necessary if students are to understand that one of the three names describes a higher order class and the other two names are members of that class.

Here's part of the introduction from Lesson 63.

a. (Display:) [63:5A]

a. boys   girls →children

- Look at the names on this number family arrow. ✔
  The small numbers are for boys and girls.
  The big number is for children.
b. Here's how you know that children is the name for the big number.
- All boys are children.
  Say the sentence. (Signal.) *All boys are children.*
- All girls are children.
  Say the sentence. (Signal.) *All girls are children.*
c. Here are the questions you ask.
- (Touch **boys.**) Are all boys children?
  Say the question for boys. (Signal.) *Are all boys children?*
- (Touch **girls.**) Say the question for girls. (Signal.) *Are all girls children?*
d. This time you'll say the questions and answer them.
- (Touch **boys.**) Say the question for boys. (Signal.) *Are all boys children?*
  What's the answer? (Signal.) *Yes.*
- (Touch **girls.**) Say the question for girls. (Signal.) *Are all girls children?*
  What's the answer? (Signal.) *Yes.*
  If all boys are children and all girls are children, the big number is children.
- How do you know that children is the big number? (Call on a student. Idea: *All boys are children, and all girls are children.*)
k. (Display:) [63:5D]

d. cows   animals →horses

This number family is silly. We'll figure out why.
- What are the names for the small numbers? (Signal.) *Cows and animals.*
  The name for the big number is horses.

l. (Point to **cows.**) Say the question for cows. (Signal.) *Are all cows horses?*
- What's the answer? (Signal.) *No.*
  So horses can't be the big number. That's silly. Cows are not horses.
m. (Point to **animals.**) Say the question for animals. (Signal.) *Are all animals horses?*
- What's the answer? (Signal.) *No.*
  So horses can't be the big number.
n. (Change to show:) [63:5E]

d. cows   horses →animals

Here's a number family with the same names in a different order.
- What are the names for the small numbers? (Signal.) *Cows and horses.*
- What's the name for the big number? (Signal.) *Animals.*
o. Raise your hand if you think this is a correct number family. ✔
- How do you know it's correct? (Call on a student. Idea: *All cows are animals, and all horses are animals.*)

from Lesson 63, Exercise 5

The test that the teacher presents for the first example is:

Are all boys children? . . . Are all girls children?

The test for the second example shown includes:

Are all cows horses? . . . Are all animals horses?

The correct family shows that all cows are animals and all horses are animals.

cows   horses →animals

Teaching Note: If students have difficulty identifying the name for the big number on later lessons, use a variation of the test presented above to show which is the big number.

Are all tools screwdrivers?

Are all screwdrivers tools?

So which is the big number, screwdrivers or tools?

Note that if you perform the test with two of the members of the family, you'll identify the big number and won't have to repeat the test with the other member of the family.

On Lesson 65, students are presented with two members of the same higher-order class. For each class, students identify the name for the big number and complete the number family so it has letters for the three names.

Here's the first part of the exercise:

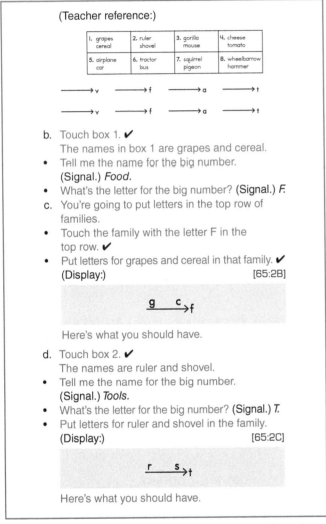

b. Touch box 1. ✔
   The names in box 1 are grapes and cereal.
• Tell me the name for the big number.
   (Signal.) *Food.*
• What's the letter for the big number? (Signal.) *F.*
c. You're going to put letters in the top row of families.
• Touch the family with the letter F in the top row. ✔
• Put letters for grapes and cereal in that family. ✔
   (Display:) [65:2B]

d. Touch box 2. ✔
   The names are ruler and shovel.
• Tell me the name for the big number.
   (Signal.) *Tools.*
• What's the letter for the big number? (Signal.) *T.*
• Put letters for ruler and shovel in the family.
   (Display:) [65:2C]

Here's what you should have.

from Lesson 65, Exercise 2

Teaching Note: Students may have trouble locating the appropriate families. If they do, say the letter for the big number of different families and direct students to touch the families. For example:

"Look at the top row of families. Touch the family with the big number of T." ✔

"Touch the family with the big number of F." ✔

"Touch the family with the big number of V." ✔

The next type of exercise begins on Lesson 69. By Lesson 70, the Textbook display presents three names for a family. Students write letters for these names in the family. Here's part of the exercise:

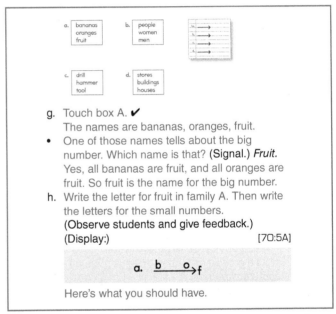

g. Touch box A. ✔
   The names are bananas, oranges, fruit.
• One of those names tells about the big number. Which name is that? (Signal.) *Fruit.*
   Yes, all bananas are fruit, and all oranges are fruit. So fruit is the name for the big number.
h. Write the letter for fruit in family A. Then write the letters for the small numbers.
   (Observe students and give feedback.)
   (Display:) [70:5A]

Here's what you should have.

from Lesson 70, Exercise 5

On Lesson 72, students work abbreviated word problems. These problems tell the numbers for two names and ask about the third.

Here's part of the exercise:

(Teacher reference:)

a. 14 girls
16 children
How many boys?

b. 9 trucks
How many vehicles?
7 cars

These problems tell names and numbers.

c. You're going to put the names in the family.
Then you'll put numbers in the family.

d. Touch problem A. ✔
I'll read the items: 14 girls. 16 children. How many boys?

• First look at the names: girls, children, boys.
One of those names is the big number.
Which name? (Signal.) *Children.*

• Put the letter for the big number in the family.
Then put in letters for the small numbers.
(Observe students and give feedback.)
(Display:) [72:6A]

Here's what you should have.

e. Now you'll put in numbers for two of those letters. Listen to the items.

f. 14 girls.

• Does that give a number for girls?
(Signal.) *Yes.*

• What number? (Signal.) *14.*

g. 16 children.

• Does that give a number for children?
(Signal.) *Yes.*

• What number? (Signal.) *16.*

h. How many boys?

• Does that give a number for boys? (Signal.) *No.*
That's the number you have to figure out.

i. Put the numbers for girls and children in the family. Then stop. ✔
(Add to show:) [72:6B]

Here's the family with numbers for girls and children.

from Lesson 72, Exercise 6

Note that the focus of this part of the exercise is not to solve the problem but to set it up properly. After the teacher has set up two problems, students work the number problems and write the answer.

Students continue to work on these abbreviated word problems through Lesson 78.

Teaching Note: Pay close attention to any mistakes students make in setting up the problems because these mistakes will persist unless they are corrected. Problems will escalate when students later work mixed sets of problems that contain more than one problem type.

On Lesson 84, students work word problems that have complete sentences. They also write the unit name in the answer.

Here's the first part of the exercise:

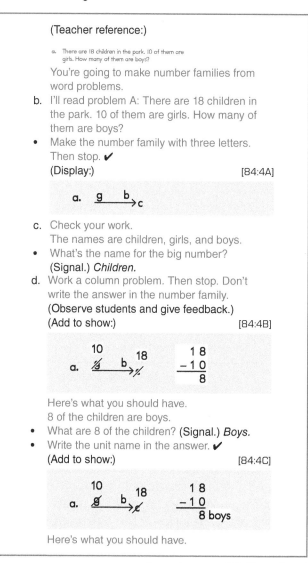

(Teacher reference:)

a. There are 18 children in the park. 10 of them are girls. How many of them are boys?

You're going to make number families from word problems.

b. I'll read problem A: There are 18 children in the park. 10 of them are girls. How many of them are boys?

• Make the number family with three letters. Then stop. ✔
(Display:) [84:4A]

c. Check your work.
The names are children, girls, and boys.

• What's the name for the big number?
(Signal.) *Children.*

d. Work a column problem. Then stop. Don't write the answer in the number family.
(Observe students and give feedback.)
(Add to show:) [84:4B]

Here's what you should have.
8 of the children are boys.

• What are 8 of the children? (Signal.) *Boys.*

• Write the unit name in the answer. ✔
(Add to show:) [84:4C]

Here's what you should have.

from Lesson 84, Exercise 4

Teaching Note: On the preceding lessons, students worked with questions. Students identified the unit name that answered the question and wrote the unit name in the answer.

Students continue to work classification problems that are structured through Lesson 87. After Lesson 87, students work separate blocks of word problems as part of their Independent Work. These are start-end problems, comparison problems, and classification problems.

Beginning on Lesson 125 and continuing through Lesson 129, students work sets of mixed problems. This work is directed by the teacher who reads each problem, directs students to make the number family with two numbers, and then directs students to work all the problems and write each answer with a unit name.

Here's the set of problems students work on Lesson 125:

---

a. Jane was 7 years older than Dan. Dan was 23 years old. How many years old was Jane?

b. There were blue trucks and red trucks on the street. 13 trucks were blue. There were 30 trucks in all. How many red trucks were there?

c. A truck started out with 50 boxes. Then workers took 19 boxes from the truck. How many boxes were still on the truck?

d. 17 boys and 26 girls were in the park. How many children were in the park?

---

from Textbook Lesson 125, Part 3

Column Addition and Subtraction (Lessons 4–90)

CMC Level C teaches basic column addition and subtraction, including carrying for addition and renaming for subtraction. The structured work with column problems starts on Lesson 4 and continues through Lesson 88.

On Lessons 4 through 10, students work addition problems that have 2-digit numbers. None of these problems requires carrying. The procedure students follow is to first say the problem for the ones column, then say the problem for the tens column, and then write answers for each column.

Here's the exercise from Lesson 5:

- Pencils down. ✔
  (Teacher reference:)　　　　　　　　　　R Part J

  a. 11　　　b. 25
  　+48　　　　+12

  c. 78　　　d. 16
  　+20　　　　+31

  These problems have two-digit numbers.
  b. Touch and read problem A. (Signal.) *11 plus 48.*
  - Say the problem for the ones column. (Signal.) *1 + 8.*
  - Say the problem for the tens column. (Signal.) *1 + 4.*
  - (Repeat until firm.)
  c. Once more: Say the problem for the ones column. (Signal.) *1 + 8.*
  - What's the answer? (Signal.) *9.*
  - Say the problem for the tens column. (Signal.) *1 + 4.*
  - What's the answer? (Signal.) *5.*
  d. Problem B. Touch and read the problem. (Signal.) *25 + 12.*
  - Say the problem for the ones column. (Signal.) *5 + 2.*
  - What's the answer? (Signal.) *7.*
  - Say the problem for the tens column. (Signal.) *2 + 1.*
  - What's the answer? (Signal.) *3.*
  e. Problem C. Touch and read the problem. (Signal.) *78 + 20.*
  - Say the problem for the ones column. (Signal.) *8 + zero.*
  - What's the answer? (Signal.) *8.*
  - Say the problem for the tens column. (Signal.) *7 + 2.*
  - What's the answer? (Signal.) *9.*
  f. Problem D. Touch and read the problem. (Signal.) *16 + 31.*
  - Say the problem for the ones column. (Signal.) *6 + 1.*
  - What's the answer? (Signal.) *7.*
  - Say the problem for the tens column. (Signal.) *1 + 3.*
  - What's the answer? (Signal.) *4.*
  g. Write answers to all the problems in part 2. (Observe students and give feedback.)
  h. Check your work.
  Read each problem and the answer.
  - Problem A. (Signal.) *11 + 48 = 59.*
  - Problem B. (Signal.) *25 + 12 = 37.*
  - Problem C. (Signal.) *78 + 20 = 98.*
  - Problem D. (Signal.) *16 + 31 = 47.*

---

from Lesson 5, Exercise 4

In steps B through F, students say the problem and the answer for each column. In step G, students write answers to all the problems. In step H, students check their work.

> **Teaching Note:** The problem that students work for each column is a fact that has been practiced in several lessons before it appears in a column problem. This pattern occurs throughout the program. First, facts are practiced in isolation. Then they are incorporated into column problems.
>
> At the beginning of this exercise, you direct students to put their pencils down. You don't want students to write anything during the presentation of steps B through F. A good plan, particularly early in the program, is to walk among students as you present Workbook exercises like this one. Make sure students touch problems before they read them and follow the other directions you present.
>
> When you observe and give feedback in step G, make sure that students are working the columns in the proper order—first the ones column and then the tens column.
>
> During the workcheck, step H, students are not to erase what they wrote. If an answer is wrong, the student should write the correct answer next to it but not erase the original answer. It's very important for you to know specifically which mistakes students make and how many students are making mistakes so you can judge whether the group is performing well enough to go to the following lesson during the next period or to review the current lesson.

Column subtraction is first introduced through the Facts track on Lesson 6. The first two-digit column subtraction is introduced on Lesson 13 with very little structure. The reason is that the subtraction procedure parallels that used for addition. Students work the problems for the ones digit first, then the column for the tens.

Here's the introduction:

| a. $\begin{array}{r} 5\ 8 \\ -1\ 2 \\ \hline \end{array}$ | b. $\begin{array}{r} 4\ 9 \\ -3\ 0 \\ \hline \end{array}$ | c. $\begin{array}{r} 6\ 5 \\ -2\ 1 \\ \hline \end{array}$ | d. $\begin{array}{r} 7\ 7 \\ -2\ 0 \\ \hline \end{array}$ |

> These are column problems that subtract. You work them the same way you work addition problems.
> - Read problem A. (Signal.) *58 – 12.*
> - Read the problem for the ones. (Signal.) *8 – 2.* What's the answer? (Signal.) *6.*
> - Read the problem for the tens. (Signal.) *5 – 1.* What's the answer? (Signal.) *4.*
> - Write the answer to the problem. ✔
> b. Check your work.
> - Everybody, read the problem and the answer. (Signal.) *58 – 12 = 46.*
> c. You'll work the rest of the problems as part of your Independent Work.

from Lesson 13, Exercise 10

If students have any trouble, repeat the same steps used for item A to work several other problems.

Students continue to work column addition and subtraction problems as Independent Work through Lesson 20.

On Lesson 21, work on carrying begins. By this lesson, students have had a lot of practice in adding three numbers.

In the first part of the exercise, the teacher asks students if different numbers have one digit or two digits. Then the teacher presents the rule for carrying: You write the tens digit in the tens column and the ones digit in the ones column. The box above the tens column prompts where the tens digit is written.

Here's part of the exercise that applies the rule:

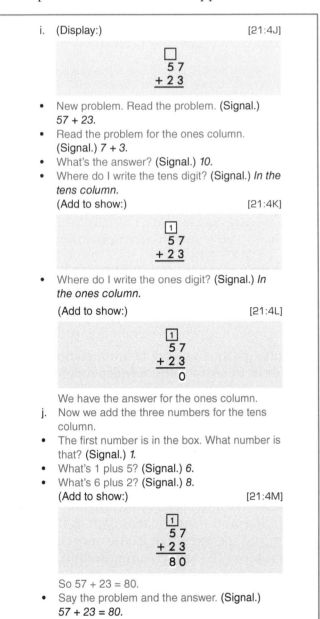

i. (Display:)   [21:4J]

$$\begin{array}{r} \square \\ 57 \\ +23 \\ \hline \end{array}$$

- New problem. Read the problem. (Signal.) *57 + 23.*
- Read the problem for the ones column. (Signal.) *7 + 3.*
- What's the answer? (Signal.) *10.*
- Where do I write the tens digit? (Signal.) *In the tens column.*
  (Add to show:)   [21:4K]

$$\begin{array}{r} \boxed{1} \\ 57 \\ +23 \\ \hline \end{array}$$

- Where do I write the ones digit? (Signal.) *In the ones column.*
  (Add to show:)   [21:4L]

$$\begin{array}{r} \boxed{1} \\ 57 \\ +23 \\ \hline 0 \end{array}$$

We have the answer for the ones column.
j. Now we add the three numbers for the tens column.
- The first number is in the box. What number is that? (Signal.) *1.*
- What's 1 plus 5? (Signal.) *6.*
- What's 6 plus 2? (Signal.) *8.*
  (Add to show:)   [21:4M]

$$\begin{array}{r} \boxed{1} \\ 57 \\ +23 \\ \hline 80 \end{array}$$

So 57 + 23 = 80.
- Say the problem and the answer. (Signal.) *57 + 23 = 80.*

from Lesson 21, Exercise 4

Teaching Note: Note that you write the digits of the carried number in the same order you write them in other contexts—the tens digit first, the ones digit next. The purpose of this convention is to demystify carrying. You're simply writing a two-digit number so that the tens digit is in the tens column and the ones digit is in the ones column. If you follow the conventions specified in the exercise, students have a far better understanding of the relationship between the number of digits and the order for writing the digits.

Students who have gone through *CMC Level B* are experienced with carrying and should find the initial exercises very easy. Treat the exercises as reviews that you go through very quickly.

By Lesson 30, students are working problems that carry with very little structure. The problems do not have a box above the tens column.

Here's the set of problems they work:

a. $\begin{array}{r} 76 \\ +19 \\ \hline \end{array}$ b. $\begin{array}{r} 69 \\ +23 \\ \hline \end{array}$ c. $\begin{array}{r} 29 \\ +52 \\ \hline \end{array}$ d. $\begin{array}{r} 48 \\ +33 \\ \hline \end{array}$

Workbook Lesson 30, Part 5

Here's the structure the teacher provides:

These problems don't have a box in the tens column.
Remember, if the answer to the problem for the ones is a two-digit number, you write the tens digit above the other tens digit.
b. Touch and read problem A. (Signal.) *76 + 19.*
- Work the problem and write the answer. (Observe students and give feedback.)
c. Check your work.
  (Display:)   [30:7A]

$$a.\quad \begin{array}{r} 1 \\ 76 \\ +19 \\ \hline 95 \end{array}$$

Here's what you should have.
- Read the problem and the answer. (Signal.) *76 + 19 = 95.*

from Lesson 30, Exercise 7

The next problem type that students work involves renaming. For example:

$$\begin{array}{r} 62 \\ -13 \\ \hline \end{array}$$

Students can't work the problem in the ones column (2 – 3). So they rewrite the top number:

$$\begin{array}{r} 5 \\ \cancel{6}^12 \\ \end{array}$$

The new problem for the ones is 12 – 3. This is a problem they can work. The new problem for the tens column is 5 – 1.

The procedure used in *CMC Level C* for teaching this operation is not strictly procedural. Rather, it is designed to show students that the original top number (62) and the rewritten number $\overset{5}{\cancel{6}}2$ are equivalent:

$$62 + 2 = 62$$
$$50 + 12 = 62$$

The program teaches this relationship through expanded notation exercises. The program refers to expanded notation as "place-value addition."

The first exercises are introduced on Lesson 2. Students say the place-value addition for regular 2-digit numbers: $70 + 5 = 75$, $90 + 2 = 92$.

The sequence introduces teen numbers after students are well practiced with the range of regular two-digit numbers. Teen numbers are delayed because there is not a close correspondence between a name like *fourteen* and the place-value addition $10 + 4$.

After students have practiced saying place-value addition equations for teen numbers ($10 + 2 = 12$, $10 + 1 = 11$), students learn to write a new place-value addition for various numbers. They subtract a ten from the value for the tens digit and add it to the ones digit:

$$40 + 3 = 43$$
$$30 + 13 = 43$$

Here's part of the exercise from Lesson 26. This is the second day that students work with the new place-value fact.

---

a. You're going to say the place-value fact for different numbers.
- Say the place-value fact for 53. (Signal.) *50 + 3 = 53.*
- Say the place-value fact for 29. (Signal.) *20 + 9 = 29.*
- Say the place-value fact for 70. (Signal.) *70 + 0 = 70.*
- Say the place-value fact for 71. (Signal.) *70 + 1 = 71.*
- Say the place-value fact for 17. (Signal.) *10 + 7 = 17.*
- (Repeat until firm.)

b. (Display:) [26:5A]

$$30 + 6 = 36$$
$$40 + 8 = 48$$
$$70 + 1 = 71$$
$$50 + 3 = 53$$

c. (Point to **30**.) You're going to subtract 10 from this number.
- What's 30 − 10? (Signal.) *20.*

d. (Point to **40**.) You're going to subtract 10 from this number.
- Say the problem. (Signal.) *40 − 10.*
- What's the answer? (Signal.) *30.*

---

e. (Point to **70**.) You're going to subtract 10 from this number.
- Say the problem. (Signal.) *70 − 10.*
- What's the answer? (Signal.) *60.*

f. (Point to **50**.) You're going to subtract 10 from this number.
- Say the problem. (Signal.) *50 − 10.*
- What's the answer? (Signal.) *40.*
- (Repeat until firm.)

g. You've subtracted 10 from the tens number. You're going to add 10 to the ones number.
- (Point to **6**.) What's 10 + 6? (Signal.) *16.*
- (Point to **8**.) What's 10 + 8? (Signal.) *18.*
- (Point to **1**.) What's 10 + 1? (Signal.) *11.*
- (Point to **3**.) What's 10 + 3? (Signal.) *13.*

h. This time you're going to subtract 10 from the tens number and add that 10 to the ones number.

i. (Point to **30**.) What's 30 − 10? (Signal.) *20.* (Change to show:) [26:5B]

$$30 + 6 = 36$$
$$20$$

- What's 10 + 6? (Signal.) *16.* (Add to show:) [26:5C]

$$30 + 6 = 36$$
$$20 + 16 = 36$$

- Read the new place-value fact for 36. (Signal.) *20 + 16 = 36.*

j. (Display:) [26:5D]

$$40 + 8 = 48$$

- (Point to **40**.) What's 40 − 10? (Signal.) *30.* (Add to show:) [26:5E]

$$40 + 8 = 48$$
$$30$$

- What's 10 + 8? (Signal.) *18.* (Add to show:) [26:5F]

$$40 + 8 = 48$$
$$30 + 18 = 48$$

- Read the new place-value fact for 48. (Signal.) *30 + 18 = 48.*

from Lesson 26, Exercise 5

In step A, students say the regular place-value fact for 53, 29, 70, 71, and 17. Students should be very solid on this step.

For the rest of the exercise segment, the teacher structures the writing of the new place-value fact for two numbers—36 and 48.

In steps C through G, students subtract ten from each of the tens and add that ten to the ones.

Starting in step H, students subtract 10 from 30 and add 10 to 6. This makes the new place-value fact $20 + 16 = 36$.

The process is repeated for 48. The new place-value fact is 30 + 18.

Teaching Note: In steps C through F, students first subtract 10 in all four examples, then add 10 in all four examples. This kind of juxtaposition makes it possible for you to identify any problems with the particular operation that is isolated—subtracting 10 or adding 10. Once the students are firm on this step, they are less likely to make mistakes when the isolated step is integrated into the other steps. Now students don't repeat the single operation but chain together two operations—subtracting and adding—to write the new place-value fact.

Before proceeding in the exercise, make sure students are firm on steps C through F (subtracting) and on step G (adding). If you have to repeat this sequence three or four times, do it. It will save time later.

Note that it is certainly possible to teach students the rewriting procedure for column problems like $\begin{smallmatrix} 3\,6 \\ -1\,8 \end{smallmatrix}$ as nothing more than rote steps: cross out the 3, write 1 less above it, and write 1 in front of the ones digit. The reason this kind of routine is not used in *CMC* is that we want to make sure students understand that they are not performing magic but are rewriting the value so when the rewriting is completed, they have exactly the same value they started with: 36.

Once students understand that the rewriting does not change the value of the number, they complete pairs of equations of the form __ + __ = 23. The top equation will show the regular place-value fact. The equation below will show the new place-value fact.

Here's part of the exercise from Lesson 29:

(Teacher reference:)   [R] [Part N]

a. ___ + ___ = 23   b. ___ + ___ = 76

new ___ + ___ = 23   new ___ + ___ = 76

You're going to complete the equations.
b. Problem A: The top equation will show the simple place-value fact for 23. The bottom equation will show the new place-value fact for 23.
- Complete the simple place-value fact for 23. ✔
- Everybody, read the place-value fact for 23. (Signal.) *20 + 3 = 23.*
c. Raise your hand when you know what number you'll write below 20. ✔
- What number? (Signal.) *10.*
  Yes, 20 – 10 = 10.
- Raise your hand when you know what number you'll write below 3. ✔
- What number? (Signal.) *13.*
  Yes, 10 + 3 is 13.
d. Remember, you subtract 10 from 20 and add that 10 to the 3.
- Complete the new place-value fact for 23. ✔
- Everybody, read the new place-value fact for 23. (Signal.) *10 + 13 = 23.*
e. Touch problem B. ✔
  The top equation will show the simple place-value fact for 76. The bottom equation will show the new place-value fact for 76.
- Complete the simple place-value fact for 76. ✔
- Everybody, read the place-value fact for 76. (Signal.) *70 + 6 = 76.*
f. Now complete the new place-value fact. Remember, you subtract 10 from the tens number and add that 10 to the ones number. ✔
- Everybody, read the new place-value fact for 76. (Signal.) *60 + 16 = 76.*

from Lesson 29, Exercise 2

Teaching Note: Follow the script. It provides a fair amount of structure for the first example, but then it fades the structure for the remaining examples. This variation is designed to make sure that students are firm and then to show them what they are expected to do. For the later examples, students are expected to work the problems without help from the teacher. If you provide structure for these problems, you're sending the wrong message to students. They are not to depend on you for detailed directions. Rather, they are to use what they know and work the problems.

Starting on Lesson 33, students rewrite numbers to show the new place-value addition. Here's the first part of the introduction:

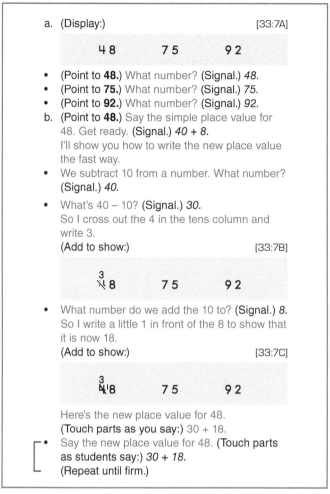

from Lesson 33, Exercise 7

The procedure shows students how to write the new place-value addition the fast way.

By Lesson 39, students rewrite numbers with little teacher direction. Here's part of the exercise from Lesson 39:

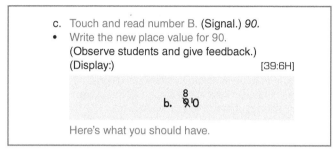

from Lesson 39, Exercise 6

Lesson 41 presents complete problems. The operation of rewriting is keyed to the ones column. Students read the problem for the ones. If the number they subtract is too large, they rewrite the top number and work the problem for the new place value. Students have already worked sets of single-digit problems that include some problems that students can work, $\overset{6}{-4}$ and some they can't work, $\overset{6}{-8}$. Students crossed out problems they can't work.

Here's part of the exercise from Lesson 43 where students are guided through the steps for working a two-digit problem that requires renaming:

a. (Display:)   [43:5A]

$$\begin{array}{r} 5\,0 \\ -1\,9 \\ \hline \end{array}$$

- Read the problem. (Signal.) *50 – 19.*
- Read the problem for the ones. (Signal.) *0 – 9.*
- Can you work that problem? (Signal.) *No.*
  So you have to rewrite a number.
- Read the number you'll rewrite. (Signal.) *50.*
b. Which digit do I cross out? (Signal.) *5.*
   (Add to show:)   [43:5B]

$$\begin{array}{r} \cancel{5}\,0 \\ -1\,9 \\ \hline \end{array}$$

- What do I write above the 5? (Signal.) *4.*
- What do I write in front of the zero? (Signal.) *1.*
  (Add to show:)   [43:5C]

$$\begin{array}{r} ^4\cancel{5}\,^10 \\ -1\,9 \\ \hline \end{array}$$

- Say the new problem for the ones. (Signal.)
  10 – 9.
- Say the new problem for the tens. (Signal.) *4 – 1.*
c. Once more: Say the new problem for the ones. (Signal.) *10 – 9.*
- What's the answer? (Signal.) *1.*
  (Add to show:)   [43:5D]

$$\begin{array}{r} ^4\cancel{5}\,^10 \\ -1\,9 \\ \hline 1 \end{array}$$

- Say the problem for the tens. (Signal.) *4 – 1.*
- What's the answer? (Signal.) *3.*
  (Add to show:)   [43:5E]

$$\begin{array}{r} ^4\cancel{5}\,^10 \\ -1\,9 \\ \hline 3\,1 \end{array}$$

I'll read the problem we started with and the answer: 50 – 19 = 31.
- Everybody, read the problem and the answer. (Signal.) *50 – 19 = 31.*
  (Repeat until firm.)

from Lesson 43, Exercise 5

Students determine that they can't work the problem for the ones, so they read the number that has to be rewritten (50) and direct the teacher to write 4 above the 5, and 1 in front of the zero. Then they say the new problems for the ones and tens.

Teaching Note: At the end of step C, students read the problem they started with and the answer. This step is important because it shows that the rewriting did not change either the value of the top number or the problem students started with. Repeat the task if student responses are weak.

Starting on Lesson 49, students work mixed sets of problems, some of which require rewriting.

Here's part of the exercise from Lesson 50:

(Teacher reference:)   R Part I

a.	b.	c.	d.	e.
6 2	6 1	7 0	9 2	9 3
– 1 1	– 1 2	– 2 3	– 8 3	– 8 2

For some of these problems, you have to write the new place value for the top number.
b. Problem A: 62 – 11.
- Read the problem for the ones. (Signal.) *2 – 1.*
- Can you work that problem? (Signal.) *Yes.*
c. Problem B: 61 – 12.
- Read the problem for the ones. (Signal.) *1 – 2.*
- Can you work that problem? (Signal.) *No.*
  So you'll have to write the new place value for the top number.
d. Problem C: 70 – 23.
- Read the problem for the ones. (Signal.) *0 – 3.*
- Can you work that problem? (Signal.) *No.*
e. Problem D: 92 – 83.
- Read the problem for the ones. (Signal.) *2 – 3.*
- Can you work that problem? (Signal.) *No.*
f. Problem E: 93 – 82.
- Read the problem for the ones. (Signal.) *3 – 2.*
- Can you work that problem? (Signal.) *Yes.*

from Lesson 50, Exercise 2

After students identify whether they can work the problem for the ones, they work the problems one at a time and check their work.

Later in the sequence, students work mixed sets of addition and subtraction problems independently and solve word problems that require carrying or renaming. On Lessons 83–90, students apply carrying and renaming to problems that have three-digit numbers.

CHECKING ANSWERS

This work starts on Lesson 96 and continues through 114.

CMC Level C introduces two strategies for students to check answers. One is to work addition problems with numbers in a different order. Another is to work the inverse operation—subtraction to check addition and addition to check subtraction.

Here's part of the exercise from Lesson 97. Students check the answer by adding the values from bottom to top.

(Teacher reference:)

a.	b.	c.	d.
2	4	8	9
4	1	4	3
+9	+9	+4	+7
16	13	16	19

These problems add three numbers. Some of the answers are wrong. You learned a way to check the answer to these problems.
You add the numbers from the bottom up. If the answer is correct, it is the same answer you get when you add the numbers from the top down.

b. Touch problem A. ✔
* What's the answer shown? (Signal.) *16.*
  You're going to check each answer by working the problem from the bottom up.
* Touch and say the first problem you work. (Signal.) *9 + 4.*
  What's the answer? (Signal.) *13.*
* Say the next problem you work. (Signal.) *13 + 2.*
  What's the answer? (Signal.) *15.*
* Is that the same answer that is shown? (Signal.) *No.*
  So 16 is wrong.
c. Let's add from the top down and make sure we end up with 15.
* Touch and say the first problem you work. (Signal.) *2 + 4.*
  What's the answer? (Signal.) *6.*
* Touch and say the next problem you work. (Signal.) *6 + 9.*
  What's the answer? (Signal.) *15.*
  So 15 is correct.
* Cross out 16 and write 15 below. ✔

from Lesson 97, Exercise 2

On Lesson 106, the inverse-operation strategy is introduced. Here's part of the exercise from Lesson 107:

(Teacher reference:)

a.	b.	c.
174	99	256
−117	+104	− 65
_____	_____	_____

b. Touch and read problem A. (Signal.) *174 − 117.*
* Work the problem. ✔
* Everybody, what's 174 − 117? (Signal.) *57.*
* On the lines below, write the two addition facts that have the same numbers.
  (Observe students and give feedback.)
  (Display:)   [107:2A]

  a. **57 + 117 = 174**

Here's one addition fact you should have.
* Say the fact. (Signal.) *57 + 117 = 174.*
* What's the other addition fact? (Signal.) *117 + 57 = 174.*
c. Touch and read problem B. (Signal.) *99 + 104.*
* Work the problem. ✔
* Everybody, what's 99 + 104? (Signal.) *203.*
* On the lines below, write the two subtraction facts that have the same numbers.
  (Observe students and give feedback.)
  (Display:)   [107:2B]

  b. **203 − 104 = 99**

Here's one subtraction fact you should have.
* Say the fact. (Signal.) *203 − 104 = 99.*
* What's the other subtraction fact? (Signal.) *203 − 99 = 104.*

from Lesson 107, Exercise 2

Teaching Note: For all problems students write two facts that have the inverse operation. Students may write them in either order. When you check students' work, you write one of the facts on the board and direct students to say the other addition fact.

Starting on Lesson 109, students check problems. Some have a wrong answer. If students identify an answer as wrong, they work the original problem to determine the right answer.

Here's part of the exercise from Lesson 109:

(Teacher reference:)

```
a.  123      b.  408      c.   58
   + 14         +152         +218
    138          560          266
```

The answers to some of these problems are wrong. You'll check by working a subtraction problem. If the answer is wrong, you'll copy the problem and figure out the correct answer.

b. Touch problem A. ✔
• Read the problem and the answer. (Signal.) *123 + 14 = 138.*
• Work a subtraction problem and see if you end up with the same three numbers that are in the addition problem.
   (Observe students and give feedback.)
• Everybody, did you end up with the same three numbers that are in the addition problem? (Signal.) *No.*
• So is the answer to the addition problem right or wrong? (Signal.) *Wrong.*
   Yes, the answer for problem A is wrong.
c. Copy the problem 123 plus 14 and work it. See if you get a different answer from the one shown in your textbook.
   Pencils down when you're finished.
   (Observe students and give feedback.)
• Everybody, what's 123 plus 14? (Signal.) *137.*

---

from Lesson 109, Exercise 5

On the following lessons, students work problems with less structure.

Mental Math (Lessons 9–123)

Exercises involving mental math start on Lesson 9 and occur throughout *CMC Level C*. The main purpose of these exercises is to promote understanding of relationships and to make this understanding strong enough that students are able to perform different mental-math operations fluently.

For instance, one group of mental-math exercises extends what students know about counting by tens and adding ten to single-digit numbers.

• Students learn to count by tens: 10, 20, 30 . . .
• Then students perform operations that relate the counting numbers to addition: What's 40 + 10? What's 70 + 10?

Later in the sequence, they do pairs of problems that extend the scope of adding tens:

• What's 40 + 10?
• So what's 42 + 10?
• What's 47 + 10?

Students also apply this generalization to subtraction:

• What's 40 − 10?
• What's 47 − 10?
• What's 247 − 10?

Later, students use what they know about adding ten to adding nine. Students learn that adding nine is one less than adding ten.

• What's 63 + 10?
• So what's 63 + 9?

Also in this track, students add multiples of ten:

• What's 2 + 5?
• So what's 20 + 50?

Students also extend what they know about single-digit addition and subtraction to two-digit numbers.

• What's 8 − 6?
• So what's 58 − 6?

Students also add and subtract hundreds numbers.

• What's 3 + 5?
• So what's 300 + 500?
• What's 900 − 100?
• So what's 952 − 100?

ADDING TEN OR NINE

Here's part of the first mental-math exercise from Lesson 9:

a. Count by tens to 100. Get ready. (Signal.) *10, 20, 30, 40, 50, 60, 70, 80, 90, 100.*
   (Repeat until firm.)
b. You'll tell me the number you get when you add 10.
   - Listen: 10. What's 10 + 10? (Signal.) *20.*
   - 20. What's 20 + 10? (Signal.) *30.*
   - 30. What's 30 + 10? (Signal.) *40.*
   - What's 40 + 10? (Signal.) *50.*
   - What's 50 + 10? (Signal.) *60.*
   (Repeat until firm.)
c. Listen: 30. What's 30 + 10? (Signal.) *40.*
   - What's 10 + 10? (Signal.) *20.*
   - What's 40 + 10? (Signal.) *50.*
   - What's 20 + 10? (Signal.) *30.*
   - What's 50 + 10? (Signal.) *60.*
   (Repeat until firm.)
d. Listen: 40. What's 40 + 10? (Signal.) *50.*
   So 41 + 10 = 51.
   - What's 41 + 10? (Signal.) *51.*
   - What's 42 + 10? (Signal.) *52.*
   - What's 46 + 10? (Signal.) *56.*
   - What's 48 + 10? (Signal.) *58.*
   (Repeat until firm.)

from Lesson 9, Exercise 1

First, students count by ten. Then they say the addition for tens numbers presented in the same order (10, 20, 30 …). In step C, the numbers are presented out of order. In step D, the numbers are from the same decade (40, 41, 42, 46 …).

Starting on Lesson 10, students relate facts that add nine to familiar facts that add ten.

Here's part of the exercise from Lesson 11:

a. (Display:) [11:8A]

3 + 10 =	3 + 9 =
7 + 10 =	7 + 9 =
9 + 10 =	9 + 9 =
5 + 10 =	5 + 9 =
8 + 10 =	8 + 9 =

These problems plus 10.
b. (Point to **3 + 10.**) Say the fact for 3 plus 10. (Signal.) *3 plus 10 equals 13.*
   - Say the fact for 7 plus 10. (Signal.) *7 + 10 = 17.*
   - Say the fact for 9 plus 10. (Signal.) *9 + 10 = 19.*
   - Say the fact for 5 plus 10. (Signal.) *5 + 10 = 15.*
   - Say the fact for 8 plus 10. (Signal.) *8 + 10 = 18.*
   (Repeat until firm.)
c. (Point to **3 + 9.**) The problems in this column do not plus 10. They plus **9.**
   So the answers are one less than the problems that plus 10.
d. (Point to **3 + 10.**) Say the fact for 3 plus 10. (Signal.) *3 + 10 = 13.*
   - So 3 plus 9 is not 13. It's **12.**
   What's 3 plus 9? (Signal.) *12.*
e. (Point to **7 + 10.**) Say the fact for 7 plus 10. (Signal.) *7 plus 10 equals 17.*
   - So 7 plus 9 is not 17. It's **16.**
   What's 7 plus 9? (Signal.) *16.*
f. This time **I'll** say the facts for plus 10. **You'll** say the facts for plus 9.
   Remember, plus 9 is one less than plus 10.
   - (Point to **3 + 10.**) 3 plus 10 equals 13.
   Say the fact for 3 plus 9. (Signal.) *3 plus 9 equals 12.*
   - (Point to **7 + 10.**) 7 + 10 = 17.
   Say the fact for 7 plus 9. (Signal.) *7 + 9 = 16.*
   - (Point to **9 + 10.**) 9 + 10 = 19.
   Say the fact for 9 plus 9. (Signal.) *9 + 9 = 18.*
   - (Point to **5 + 10.**) 5 + 10 = 15.
   Say the fact for 5 plus 9. (Signal.) *5 + 9 + 14.*
   - (Point to **8 + 10.**) 8 + 10 = 18.
   Say the fact for 8 plus 9. (Signal.) *8 + 9 = 17.*
   (Repeat until firm.)

from Lesson 11, Exercise 8

Students first say the facts that plus ten. Then the teacher introduces the relationship for facts that plus nine. The answer is one less than the answer to plus-ten facts. In step F, the teacher says the plus-ten facts, and students say corresponding plus-nine facts.

The exercise that follows on Lesson 11 requires students to write answers to paired problems—a plus-ten problem and a corresponding plus-nine problem.

Here's the set of problems they work:

a. $7 + 10 = \_\_\_$ $7 + 9 = \_\_\_$
b. $4 + 10 = \_\_\_$ $4 + 9 = \_\_\_$
c. $9 + 10 = \_\_\_$ $9 + 9 = \_\_\_$
d. $5 + 10 = \_\_\_$ $5 + 9 = \_\_\_$
e. $3 + 10 = \_\_\_$ $3 + 9 = \_\_\_$

Workbook Lesson 11, Part 4

Teaching Note: Students typically require considerable practice to learn this relationship. If they make mistakes or take a lot of time to work the problems, remind them of the rule that plus nine is one less than plus ten.

Starting on Lesson 58, students work problems that add nine to two-digit numbers.

Here's the first exercise:

a. I'll say numbers. You'll say the numbers that are one less.
• Listen: 34. What number is 1 less than 34? (Signal.) *33.*
• Listen: 78. What number is 1 less than 78? (Signal.) *77.*
• Listen: 45. What number is 1 less than 45? (Signal.) *44.*
• Listen: 29. What number is 1 less than 29? (Signal.) *28.*
• Listen: 61. What number is 1 less than 61? (Signal.) *60.*
  (Repeat until firm.)
b. You're going to add 10 and then add 9. Remember, the answer for plus 9 is one less than the answer for plus 10.
• Listen: 24 + 10. What's the answer? (Signal.) *34.* 24 + **10** is 34. So what's 24 + **9**? (Signal.) *33.*
• Listen: 68 + 10. What's the answer? (Signal.) *78.* 68 + **10** is 78. So what's 68 + **9**? (Signal.) *77.*
• Listen: 35 + 10. What's the answer? (Signal.) *45.* 35 + **10** is 45. So what's 35 + **9**? (Signal.) *44.*
• Listen: 19 + 10. What's the answer? (Signal.) *29.* 19 + **10** is 29. So what's 19 + **9**? (Signal.) *28.*
• Listen: 51 + 10. What's the answer? (Signal.) *61.* 51 + **10** is 61. So what's 51 + **9**? (Signal.) *60.*
  (Repeat until firm.)

Lesson 58, Exercise 6

In step A, students identify numbers that are one less than various two-digit numbers. In step B, they answer pairs of facts—first the plus-ten fact and then the corresponding plus-nine fact.

By Lesson 67, students write answers to plus-nine problems that are not paired with corresponding plus-ten problems.

Students learn another relationship with plus-ten facts starting on Lesson 33. The relationship is that 76 + 10 has the same answer as 10 + 76. Students are well practiced with the commutative property as it applies to facts such as 5 + 3 and 3 + 5. The exercises involving 10+ extend students' understanding of the commutative property.

Here's the exercise from Lesson 33:

a. You're going to add 10 to different numbers.
b. What's 45 + 10? (Signal.) *55.* So what's 10 + 45? (Signal.) *55.*
• What's 10 + 75? (Signal.) *85.*
• What's 10 + 78? (Signal.) *88.*
• What's 10 + 51? (Signal.) *61.*
• What's 10 + 36? (Signal.) *46.*
c. What's 38 + 10? (Signal.) *48.*
• What's 15 + 10? (Signal.) *25.*
• What's 9 + 10? (Signal.) *19.*
• What's 29 + 10? (Signal.) *39.*
• What's 10 + 43? (Signal.) *53.*
  (Repeat until firm.)

Lesson 33, Exercise 6

Note that in some of these problems, ten is the second addend. It's not always the first addend.

Later exercises extend practice to three-digit numbers, starting on Lesson 35.

Here's the exercise from Lesson 36:

a. You're going to add 10 to different numbers.
b. Listen: What's 10 + 54? (Signal.) *64.*
   So what's 10 + 154? (Signal.) *164.*
   • What's 10 + 127? (Signal.) *137.*
   • What's 10 + 125? (Signal.) *135.*
   • What's 10 + 153? (Signal.) *163.*
   • What's 10 + 118? (Signal.) *128.*
   (Repeat until firm.)
c. Listen: What's 39 + 10? (Signal.) *49.*
   • What's 135 + 10? (Signal.) *145.*
   • What's 188 + 10? (Signal.) *198.*
   (Repeat until firm.)

Lesson 36, Exercise 6

Teaching Note: If students have trouble, present a two-digit problem related to the three-digit problem they missed. Then return to the three-digit problem. For example, students have trouble with 20 + 118:

"Listen: What's 20 + 18?"

"So what's 20 + 118?"

(Repeat until firm.)

Then return to the first item in the set, and repeat the items in order.

ADDING OR SUBTRACTING MULTIPLES OF TEN

Starting on Lesson 71, students learn to expand single-digit facts into facts that are multiples of ten.

• 2 + 3 = 5.
• So 20 + 30 = 50.

Here's the exercise from Lesson 72:

a. (Display:) [72:5A]

$$20 + 40 =$$
$$60 + 30 =$$
$$70 + 10 =$$
$$20 + 60 =$$

For each problem, first you'll say the fact for the tens digits. Then you'll say the fact for the whole problem.
b. (Point to **20 + 40**.) My turn to say the fact for the 10s digits: 2 + 4 = 6.
   • Your turn: Say the fact for the tens digits. (Signal.) *2 + 4 = 6.*
   • Say the fact for 20 + 40. (Signal.) *20 + 40 = 60.*
   (Repeat until firm.)
c. (Point to **60 + 30**.) Say the fact for the tens digits. (Signal.) *6 + 3 = 9.*
   • Say the fact for 60 + 30. (Signal.) *60 + 30 = 90.*
d. (Point to **70 + 10**.) Say the fact for the tens digits. (Signal.) *7 + 1 = 8.*
   • Say the fact for 70 + 10. (Signal.) *70 + 10 = 80.*
e. (Point to **20 + 60**.) Say the fact for the tens digits. (Signal.) *2 + 6 = 8.*
   • Say the fact for 20 + 60. (Signal.) *20 + 60 = 80.*

from Lesson 72, Exercise 5

Teaching Note: If students later have trouble with problems that have multiples of ten, follow the procedure presented in this exercise. Tell students:

"Say the fact for the tens digits."

"Say the fact for" [the original problem].

Later exercises expand the range of multiples of ten to 3-digit problems: 180 + 30.

Other exercises show relationships with subtraction.

• What's 9 – 3?
• So what's 90 – 30?

OTHER FACT EXTENSIONS

Students also work mental math problems involving hundreds. They extend their knowledge of familiar facts such as 2 + 3 = 5 or 8 − 4 = 4.

- What's 2 hundred plus 3 hundred?
- What's 8 hundred minus 4 hundred?

The final relationship presented as mental math involves adding single digits to two-digit numbers, for example: 63 + 4.

Students first work the problem for the ones digits: 3 + 4 = 7. Then they work the original problem: 63 + 4 = 67.

Students also apply this relationship to subtraction.

Here's part of the exercise from Lesson 119:

---

a. (Display:)   [119:1A]

$$35 + 3$$
$$76 - 4$$
$$81 + 8$$
$$43 + 6$$

These problems are easy to work.
b. (Point to **35 + 3**.) Read this problem. (Signal.) *35 + 3.*
  - What's 5 + 3? (Signal.) *8.*
  - So what's 35 + 3? (Signal.) *38.*
  (Repeat until firm.)
c. (Point to **76 − 4**.) Read this problem. (Signal.) *76 − 4.*
  - What's 6 − 4? (Signal.) *2.*
  - So what's 76 − 4? (Signal.) *72.*
  (Repeat until firm.)
d. (Point to **81 + 8**.) Read this problem. (Signal.) *81 + 8.*
  - What's 1 + 8? (Signal.) *9.*
  - So what's 81 + 8? (Signal.) *89.*
  (Repeat until firm.)
e. (Point to **43 + 6**.) Read this problem. (Signal.) *43 + 6.*
  - What's 3 + 6? (Signal.) *9.*
  - So what's 43 + 6? (Signal.) *49.*
  (Repeat until firm.)
f. Remember, say the fact for the ones digit, and then it's easy to figure out the answer to the whole problem.

Lesson 119, Exercise 1

The discussion provided here does not cover all the relationships students learn. The exercises presented on different lessons often present mental-math examples that are relevant to what students are doing in some of the other tracks. For example, students work mental math problems that have missing addends.

Money (Lessons 14–111)

The Money track starts on Lesson 14 and continues intermittently through Lesson 111. Students first learn to identify the value of coins— pennies, nickels, dimes, and quarters. They also learn that a dollar is 100 cents. Students count the cents for combinations of coins (counting by combinations of ones, fives, tens, and 25s). They also count combinations that have bills and coins. This work is coordinated with the counting track.

Students add and subtract column problems that have dollars-and-cents values. They apply conventions for making the dollar sign and decimal point in the answer.

Students later learn to convert cent values (such as 411 cents) into dollars-and-cents notation. Finally, students solve word problems involving dollars-and-cents values. For most of these problems, students make number families. For some, they work problems that require two steps—figuring out how much a person spends to purchase specific items and how much the person has after the purchases.

Counting like-coins starts on Lesson 14. Here's the first part of the exercise from Lesson 16:

a. (Display:)   [16:9A]

This shows pennies, nickels, and dimes.

b. (Touch the row of **pennies.**) What are these coins? (Signal.) *Pennies.*
- What is each penny worth? (Signal.) *1 cent.*
- So you count by 1 for each penny. What do you count by? (Signal.) *1.*

c. (Touch the row of **dimes.**) What are these coins? (Signal.) *Dimes.*
- What is each dime worth? (Signal.) *10 cents.*
- So what do you count by for each dime? (Signal.) *10.*

d. (Touch the row of **nickels.**) What are these coins? (Signal.) *Nickels.*
- What is each nickel worth? (Signal.) *5 cents.*
- So what do you count by for each nickel? (Signal.) *5.*
  (Repeat until firm.)

e. (Touch the row of **pennies.**) What do you count by for each penny? (Signal.) *1.*
- Count by 1 for each penny. (Touch each penny as students count:) *1, 2, 3, 4, 5, 6.*
- So how much is this row worth? (Signal.) *6 cents.*
  Yes, 6 cents.

f. (Touch the row of **dimes.**) What do you count by for each dime? (Signal.) *10.*
- Count by 10 for each dime. (Touch each dime as students count:) *10, 20, 30, 40, 50, 60.*
- So how much is this row worth? (Signal.) *60 cents.*
  Yes, 60 cents.

g. (Touch the row of **nickels.**) What do you count by for each nickel? (Signal.) *5.*
- Count by 5 for each nickel. (Touch each nickel as students count:) *5, 10, 15, 20, 25, 30.*
- So how much is this row worth? (Signal.) *30 cents.*

from Lesson 16, Exercise 9

In this part of the exercise, students indicate what each type of coin is worth and then count by that number to figure out the number of cents in the row.

In the Workbook activity that follows in this exercise, students count cents for three rows of coins and write the number of cents at the end of each row.

Teaching Note: Students should not have trouble with this exercise if they are well practiced in counting by 5 and 10. When they count in steps E through G, make sure they count together, as you touch the coins. Also make sure that they touch the appropriate coins as they count cents in their Workbook.

On Lesson 17, students count rows that have more than one type of coin. Here's part of the exercise from Lesson 21:

d. (Display:)   [21:3A]

- (Point to **dimes.**) What do you count by for dimes? (Signal.) *Tens.*
- (Point to **nickel.**) What do you count by for nickels? (Signal.) *Fives.*

e. My turn to count for the dimes and the nickels. Count for the dimes. 10, 20. Now I keep counting for the nickel. 25.

f. Your turn: Count for the dimes. (Signal.) *10, 20.*
- Now keep counting for the nickel. (Signal.) *25.*
- How many cents are in the row? (Signal.) *25.* (Repeat until firm.)

g. (Display:)   [21:3B]

- (Point to **nickels.**) What do you count by for nickels? (Signal.) *Fives.*
- (Point to **penny.**) What do you count by for pennies? (Signal.) *Ones.*

h. (Point to **nickels.**) Count for the nickels. (Signal.) *5, 10, 15.*
  How many cents do you have? (Signal.) *15.*
- Now keep counting for the penny. (Signal.) *16.*
  How many cents are in the row? (Signal.) *16.*

from Lesson 21, Exercise 3

Teaching Note: For each row, students identify the numbers they'll count by, and then they count. The critical step is the transition from counting by one number and counting on by another number. In step F, students count by 10 and then by 5. Repeat this step until students are firm. When you present the other rows of coins, repeat the corresponding steps. Then return to step D and present the original sequence. Note any problems that students have. If they have trouble, plan to repeat this part of the exercise at the beginning of the next math period.

Starting on Lesson 25, students work with rows that have bills and coins. Here are the rows presented on Lesson 26:

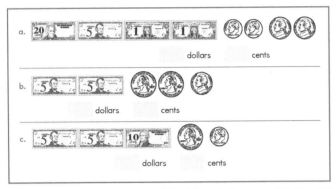

Workbook Lesson 26, Part 3

Students first count to determine the number of dollars. For the first row, students start at 20, then count by fives one time and then count by ones two times. Then they count to determine the number of cents.

Teaching Note: If students have trouble, model the counting, and then repeat the part of the exercise that directs the counting. After students have counted for all the rows, repeat the counting for the rows without first modeling.

The dollar sign and decimal point are introduced on Lesson 57. Students read dollars-and-cents values. Here's part of the exercise from Lesson 58:

a. (Display:)   [58:5A]

$6.15 $11.20 $1.05 $23.99 $50.00

b. (Point to **$**.) You learned the name of this sign.
- What's this sign? (Signal.) *A dollar sign.*
  Yes, a dollar sign.
c. (Point to **dot**.) You learned that the number that comes **before** the dot shows dollars. The number that comes **after** the dot shows cents.
- What does the number that comes before the dot show? (Signal.) *Dollars.*
- What does the number after the dot show? (Signal.) *Cents.*
d. My turn to read these dollars-and-cents numbers.
  (Point to **$6.15**.) 6 dollars and 15 cents.
  (Point to **$11.20**.) 11 dollars and 20 cents.
  (Point to **$1.05**.) 1 dollar and 5 cents.
  (Point to **$23.99**.) 23 dollars and 99 cents.
  (Point to **$50.00**.) 50 dollars.
e. Your turn to read these numbers.
- (Point to **$6.15**.) Get ready. (Touch.) *6 dollars and 15 cents.*
- (Point to **$11.20**.) Get ready. (Touch.) *11 dollars and 20 cents.*
- (Point to **$1.05**.) Get ready. (Touch.) *1 dollar and 5 cents.*
- (Point to **$23.99**.) Get ready. (Touch.) *23 dollars and 99 cents.*
- (Point to **$50.00**.) Get ready. (Touch.) *50 dollars.*
  (Repeat until firm.)

from Lesson 58, Exercise 5

Steps B through D review the dollar sign and the decimal point. Then the teacher reads the dollars-and-cents values (step D). Then students read the values (step E).

            Connecting Math Concepts

After step E, students read and copy dollar amounts.

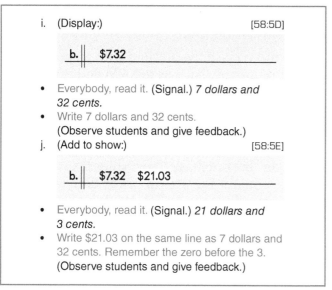

i. (Display:) [58:5D]

b. $7.32

- Everybody, read it. (Signal.) *7 dollars and 32 cents.*
- Write 7 dollars and 32 cents.
  (Observe students and give feedback.)

j. (Add to show:) [58:5E]

b. $7.32 $21.03

- Everybody, read it. (Signal.) *21 dollars and 3 cents.*
- Write $21.03 on the same line as 7 dollars and 32 cents. Remember the zero before the 3.
  (Observe students and give feedback.)

from Lesson 58, Exercise 5

Teaching Note: If students have trouble reading the amounts in step E, remind them:

"You say the number before the dot. Then say dollars."

(Point to dollars.) "What's the number before the dot?"

"What do you say after the number?"

(Point to cents.) "What's the number after the dot?"

"What do you say after the number?"

(Point.) "Read the dollars-and-cents number."

On the next three lessons, the teacher dictates dollars-and-cents amounts and students write them.

Students work column problems with dollars-and-cents values starting on Lesson 92. The problems that students work are in their Workbook. Students first write the dollar sign and the dot in the answer. Then they work the problem. The columns are referred to as the one-cent column, the ten-cent column, and the dollar column.

Here's part of the exercise from Lesson 92.

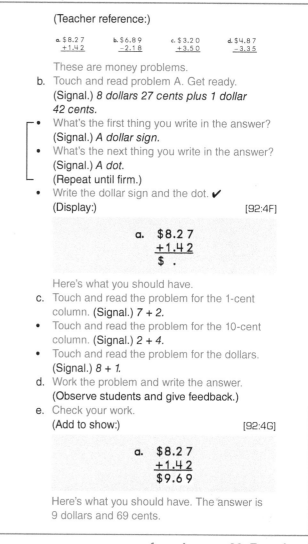

(Teacher reference:)

a. $8.27   b. $6.89   c. $3.20   d. $4.87
   +1.42   −2.18   +3.50   −3.35

These are money problems.

b. Touch and read problem A. Get ready.
   (Signal.) *8 dollars 27 cents plus 1 dollar 42 cents.*
   - What's the first thing you write in the answer? (Signal.) *A dollar sign.*
   - What's the next thing you write in the answer? (Signal.) *A dot.*
   (Repeat until firm.)
   - Write the dollar sign and the dot. ✔
   (Display:) [92:4F]

a. $8.27
   +1.42
   $.

Here's what you should have.

c. Touch and read the problem for the 1-cent column. (Signal.) *7 + 2.*
   - Touch and read the problem for the 10-cent column. (Signal.) *2 + 4.*
   - Touch and read the problem for the dollars. (Signal.) *8 + 1.*

d. Work the problem and write the answer. (Observe students and give feedback.)

e. Check your work.
   (Add to show:) [92:4G]

a. $8.27
   +1.42
   $9.69

Here's what you should have. The answer is 9 dollars and 69 cents.

from Lesson 92, Exercise 4

On Lesson 97, students work multiplication problems based on questions of the form: **How many cents is ___ nickels?**

Starting on Lesson 100, students convert large cents values into dollars-and-cents values. They apply the rule that one dollar equals 100 cents.

Here's part of the exercise:

a. You know that 100 cents equals 1 dollar. Any cents number that is more than 100 is more than 1 dollar.

b. My turn: What's 103 cents? 1 dollar and 3 cents. What's 124 cents? 1 dollar and 24 cents.
  • Your turn: What's 130 cents? (Signal.) *1 dollar and 30 cents.*
  • What's 124 cents? (Signal.) *1 dollar and 24 cents.*
  • What's 176 cents? (Signal.) *1 dollar and 76 cents.*
  • What's 198 cents? (Signal.) *1 dollar and 98 cents.*
  (Repeat until firm.)

c. My turn: What's 509 cents? 5 dollars and 9 cents.
  • Your turn: What's 524 cents? (Signal.) *5 dollars and 24 cents.*
  • What's 580 cents? (Signal.) *5 dollars and 80 cents.*
  • What's 750 cents? (Signal.) *7 dollars and 50 cents.*
  • What's 309 cents? (Signal.) *3 dollars and 9 cents.*
  (Repeat until firm.)

d. (Display:)   [100:3A]

      309 cents =

My turn to complete the equation.
  • What does 309 cents equal? (Signal.) *3 dollars and 9 cents.*
  So I write 3 dollars and 9 cents.
  (Add to show:)   [100:3B]

      309 cents = $3.09

from Lesson 100, Exercise 3

Teaching Note: In the part of the exercise that follows, students complete equations in their Workbook of the form: **803 cents = ___**. Make sure that students are firm on the verbal part of the exercise in steps A through C. If they are, completing the equations is easy.

Word problems involving dollars-and-cents amounts begin on Lesson 100 and continue through Lesson 111. Students work three kinds of problems. For the first type, students add to find the total cost of purchases. A display shows prices of different items. The problem indicates which of the items a person purchases and asks how much money the person needs.

The second type shows a wallet with a specified amount of money (such as $68.75) and asks how much the person has left after purchasing a specified item shown in the display.

The third type requires students to add and subtract. First, students calculate the total cost of more than one purchase. Then students subtract from the amount the person has to determine how much money the person has left.

Here's an example of the first type from Lesson 100:

g. I'll read problem A: A person buys the book and the hat.
   You're going to figure out how much the person needs to buy the book and the hat. So you add the numbers for the book and the hat.
  • Write the problem for A. Pencils down when you've done that much.
   (Observe students and give feedback.)
   (Display:)   [100:5A]

      a. $ 9.5 0
          +1 6.3 0

Here's what you should have.
h. Work the problem and write the answer.
   (Observe students and give feedback.)
   (Add to show:)   [100:5B]

      a. $ 9.5 0
          +1 6.3 0
          $2 5.8 0

Here's what you should have.
$9.50 + $16.30 = $25.80. So the person needs $25.80 to buy the book and the hat.

from Lesson 100, Exercise 5

Teaching Note: To work this problem, students use the various skills that have been taught earlier—aligning the columns, writing the dollar sign and the decimal point, and adding. If the earlier teaching was solid, students should have no trouble with these problems.

The second type of money problem begins on Lesson 104.

Here's part of the exercise from Lesson 105:

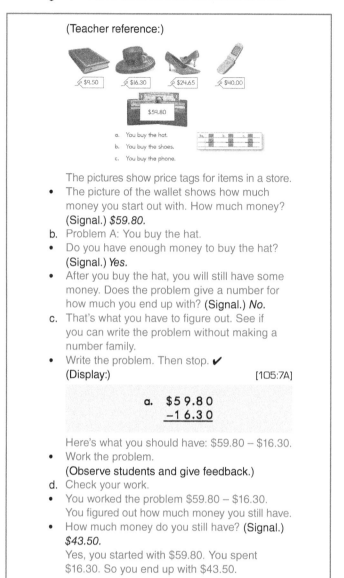

from Lesson 105, Exercise 7

In preparation for the third type of problem, students work problems in which they determine if a person has enough money to purchase specified items. Students add the costs of the items and compare the total to the amount of money the person has.

Here's part of an exercise that presents this problem format.

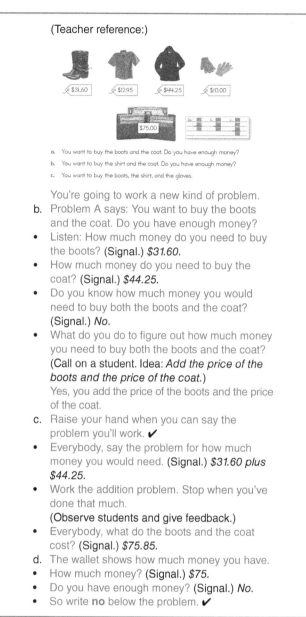

from Lesson 107, Exercise 6

For the final problem type, students make the determination of whether the person has enough money. If the answer is "yes," they figure out how much money the person will have after the purchase.

Here's part of the exercise from Lesson 111.

f. Problem B: You want to buy the book and the paper.
• Work the problem. Figure out how much you need to buy the book and the paper.
   (Observe students and give feedback.)
g. Check your work.
    (Display:)   [111:4B]

```
    b.  $ 9.2 5
       +  3.4 9
        $1 2.7 4
```

• What's the answer to the problem? (Signal.) *$12.74.*
• How much money do you need to buy the book and the paper? (Signal.) *$12.74.*
h. Now look at the wallet. Write **yes** if you have enough money. Write **no** if you don't have enough money. ✔
• Do you have enough money? (Signal.) *Yes.* Yes, you have enough money.
• Figure out how much money you still have after you buy the items.
   (Observe students and give feedback.)
i. Check your work.
    (Add to show:)   [111:4C]

```
    b.  $ 9.2 5      $2 6.7 5
       +  3.4 9     –1 2.7 4
        $1 2.7 4     $1 4.0 1
           yes
```

Here's what you should have.
• How much money do you need? (Signal.) *$12.74.*
• How much money do you have after you buy the items? (Signal.) *$14.01.*
j. Problem C: You want to buy the phone and the paper.
• Work the problem. Figure out how much money you need to buy the phone and the paper. Pencils down when you've done that much.
   (Observe students and give feedback.)

k. Check your work.
    (Display:)   [111:4D]

```
    c.  $2 5.5 0
       +  3.4 9
        $2 8.9 9
```

Here's what you should have.
• How much money did you need? (Signal.) *$28.99.*
l. Now look at the wallet. Write **yes** if you have enough money. Write **no** if you don't have enough money. ✔
• Did you have enough money? (Signal.) *No.*

from Lesson 111, Exercise 4

Multiplication (Lessons 32–124)

The Multiplication track starts on Lesson 32 and continues through Lesson 124. The basic procedure students learn is to translate multiplication problems into counting by different numbers so many times; for example, **2 × 5 = ___**.

Students translate this problem into "Count by 2, five times." To show that they count five times, they make five dots under the 5.

$$2 \times \underset{\cdot\,\cdot\,\cdot\,\cdot\,\cdot}{5} = \underline{\quad}$$

They then count by *two* for each dot: 2, 4, 6, 8, 10.

Here's the introduction from Lesson 33.

---

a. (Display:) [33:3A]

    5 × 8

- (Point to **x**.) This is a count-by sign.
- (Point to **5**.) This tells you to count by 5. What does it tell you? (Signal.) *Count by 5.*
- (Point to **8**.) This tells how many times you count. How many times? (Signal.) *8.*
- (Point to problem.) Yes, this problem tells you to count by 5 eight times. What does this problem tell you to do? (Signal.) *Count by 5 eight times.*

b. (Display:) [33:3B]

    10 × 2

- What does this problem tell you to do? (Signal.) *Count by 10 two times.* (Display:) [33:3C]

    1 × 12

- What does this problem tell you to do? (Signal.) *Count by 1 twelve times.* (Display:) [33:3D]

    2 × 4

- What does this problem tell you to do? (Signal.) *Count by 2 four times.*

c. I'll show you how to work the problem. You're going to count four times. So I make four dots under the 4. (Add to show:) [33:3E]

    2 × 4̤

I'll count by 2 four times and touch a dot each time I count. (Touch dots as you count:) 2, 4, 6, 8.
- What did I end up with? (Signal.) *8.* (Add to show:) [33:3F]

    2 × 4̤ = 8

Yes, if you count by 2 four times, you end up with 8.

---

d. (Display:) [33:3G]

    10 × 6

New problem.
- What does this problem tell you to do? (Signal.) *Count by 10 six times.* (Repeat until firm.)
- How many times does it tell you to count? (Signal.) *6.*
- So how many dots do I make under the 6? (Signal.) *Six.* (Add to show:) [33:3H]

    10 × 6̤

- What do you count by? (Signal.) *10.* I'll touch the dots. You'll count by 10 six times. (Touch dots as students count:) *10, 20, 30, 40, 50, 60.*
- How many did you end up with? (Signal.) *60.* (Add to show:) [33:3I]

    10 × 6̤ = 60

from Lesson 33, Exercise 3

Teaching Note: Make sure that students are very solid in translating what the problem tells them to do. This translation will be practiced in the following lessons, but it's important for students to become familiar with the verbal routine early.

On Lesson 37, students learn that the problem tells them what to count by, but that they read the problem another way.

Here's part of the introduction.

a. (Display:) [37:1A]

5 x 8

10 x 2

4 x 7

2 x 6

9 x 5

- (Point to **5 x 8**.) What does this problem tell you to do? (Signal.) *Count by 5 eight times.*
- (Point to **10 x 2**.) What does this problem tell you to do? (Signal.) *Count by 10 two times.*
- (Point to **4 x 7**.) What does this problem tell you to do? (Signal.) *Count by 4 seven times.*
- (Point to **2 x 6**.) What does this problem tell you to do? (Signal.) *Count by 2 six times.*
- (Point to **9 x 5**.) What does this problem tell you to do? (Signal.) *Count by 9 five times.*

b. I will show you how to read these problems. (Touch the **x** in 5 x 8.) This is called a times sign.
- What is it called? (Signal.) *A times sign.*
  I'll read the problem. (Point to symbols as you say:) *5 times 8.*
- Your turn: Read the problem. (Signal.) *5 times 8.*

c. (Point to **10 x 2**.) Read this problem. (Signal.) *10 times 2.*
- (Point to **4 x 7**.) Read this problem. (Signal.) *4 times 7.*
- (Point to **2 x 6**.) Read this problem. (Signal.) *2 times 6.*
- (Point to **9 x 5**.) Read this problem. (Signal.) *9 times 5.*

d. (Point to **x** in 5 x 8.) Remember, this is a times sign. It tells you to count by.
- What does it tell you to do? (Signal.) *Count by.*
- What does this problem tell you to do? (Signal.) *Count by 5 eight times.*

from Lesson 37, Exercise 1

In step A, students indicate what each problem tells them to do. In steps B and C, students read the problems.

In the following lessons, students work from directions that are abbreviated; however, they continue to read problems and translate them into counting operations. They work problems that involve all the numbers they count by—2, 4, 5, 9, and 10.

On Lesson 44, students apply the multiplication analysis to finding the area of rectangles. (The term *area* is not used in the early exercises.)

a. (Display:) [44:5A]

You're going to figure out how many squares there are. But you won't count all of them. You'll work a count-by problem.

b. Count the squares in the top row. Get ready. (Touch as students count:) *1, 2, 3, 4, 5.*
- How many squares are in the top row? (Signal.) *5.*
- So how many squares are in the next row? (Signal.) *5.*
- How many squares are in the bottom row? (Signal.) *5.*

c. Listen: There are 5 squares in each row. So you count by 5.
- What do you count by? (Signal.) *5.*
  (Add to show:) [44:5B]

5 x

Yes, you count by 5.

d. Listen: You count by 5 for each row because each row has 5 squares.
  Once more: You count by 5 for each row.
- How many rows are there? (Signal.) *3.*
- So how many times do you count by 5? (Signal.) *3.*
  Yes, count by 5 three times.
- Say that. (Signal.) *Count by 5 three times.*
  (Add to show:) [44:5C]

5 x 3

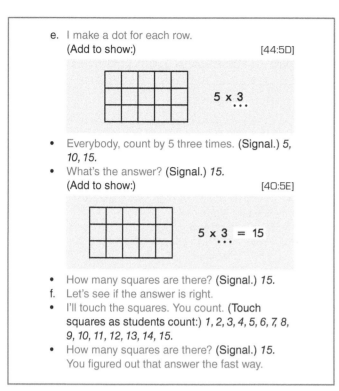

e. I make a dot for each row.
   (Add to show:)   [44:5D]

5 x 3
• • •

• Everybody, count by 5 three times. (Signal.) *5, 10, 15.*
• What's the answer? (Signal.) *15.*
  (Add to show:)   [40:5E]

5 x 3 = 15
• • •

• How many squares are there? (Signal.) *15.*
f. Let's see if the answer is right.
• I'll touch the squares. You count. (Touch squares as students count:) *1, 2, 3, 4, 5, 6, 7, 8, 9, 10, 11, 12, 13, 14, 15.*
• How many squares are there? (Signal.) *15.* You figured out that answer the fast way.

from Lesson 44, Exercise 5

Working these problems is a simple extension of making dots to work the multiplication problems. Students determine the number in each row. They make dots for the number of rows. Then they work the problem.

Starting on Lesson 80, students write multiplication problems from rows of like coins.

Here's part of the exercise from Lesson 82:

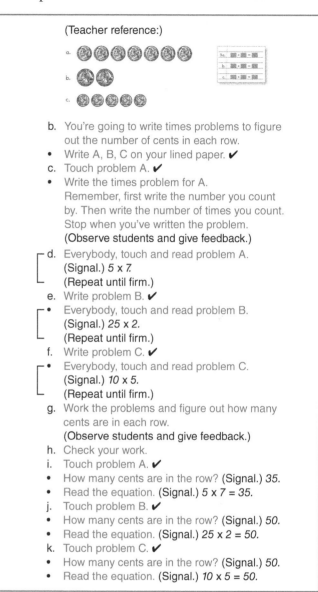

(Teacher reference:)

b. You're going to write times problems to figure out the number of cents in each row.
• Write A, B, C on your lined paper. ✔
c. Touch problem A. ✔
• Write the times problem for A. Remember, first write the number you count by. Then write the number of times you count. Stop when you've written the problem. (Observe students and give feedback.)
d. Everybody, touch and read problem A. (Signal.) *5 x 7.* (Repeat until firm.)
e. Write problem B. ✔
• Everybody, touch and read problem B. (Signal.) *25 x 2.* (Repeat until firm.)
f. Write problem C. ✔
• Everybody, touch and read problem C. (Signal.) *10 x 5.* (Repeat until firm.)
g. Work the problems and figure out how many cents are in each row. (Observe students and give feedback.)
h. Check your work.
i. Touch problem A. ✔
• How many cents are in the row? (Signal.) *35.*
• Read the equation. (Signal.) *5 x 7 = 35.*
j. Touch problem B. ✔
• How many cents are in the row? (Signal.) *50.*
• Read the equation. (Signal.) *25 x 2 = 50.*
k. Touch problem C. ✔
• How many cents are in the row? (Signal.) *50.*
• Read the equation. (Signal.) *10 x 5 = 50.*

Lesson 82, Exercise 7

Students first write the problems for the coins and then work them. Students have learned that the first number in each problem is the number of cents for the coin. The second number is the number of coins in the row.

Starting on Lesson 97, students work from questions that ask about groups of like coins. For example: "How many cents is eight nickels?" and "How many cents is four dimes?"

Here's part of the introduction from Lesson 97:

a. (Display:)   [97:4A]

> **How many cents is 8 nickels?**

- Here's a new kind of problem: How many cents is 8 nickels? Say the problem. (Signal.) *How many cents is 8 nickels?*

b. Here's how we work the problem. We first write the number of cents for each nickel.
- How many cents is each nickel? (Signal.) *5.* So we count by fives.
(Add to show:)   [97:4B]

> **How many cents is 8 nickels?**
>
> 5

c. Now we write the number of nickels we have.
- What's the number of nickels? (Signal.) *8.*
(Add to show:)   [97:4C]

> **How many cents is 8 nickels?**
>
> 5 x 8

So we count by fives 8 times.
- How many dots do I make? (Signal.) *8.*
(Add to show:)   [97:4D]

> **How many cents is 8 nickels?**
>
> 5 x 8
>   ·······

d. I'll touch the dots. You count by fives. Get ready. (Touch as students count:) *5, 10, 15, 20, 25, 30, 35, 40.*
- How many cents is 8 nickels? (Signal.) *40.*
(Add to show:)   [97:4E]

> **How many cents is 8 nickels?**
>
> 5 x 8 = 40
>   ·······

The problem asks how many cents.
- So what's the unit name in the answer? (Signal.) *Cents.*
(Add to show:)   [97:4F]

> **How many cents is 8 nickels?**
>
> 5 x 8 = 40 cents
>   ·······

from Lesson 97, Exercise 4

On Lesson 90, students learn to work multiplication problems that have a missing factor: 4 × __ = 20; 2 × __ = 6.

Before these problems are introduced, students practice making dots as they count by different numbers. They make a dot each time they count. They then count the dots to determine how many times they counted.

This component is embedded in the operation they learn for finding the missing factor.

Here's part of the exercise from Lesson 91:

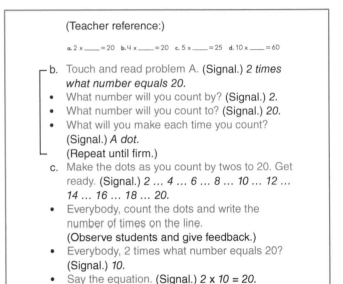

(Teacher reference:)

a. 2 x ____ = 20   b. 4 x ____ = 20   c. 5 x ____ = 25   d. 10 x ____ = 60

b. Touch and read problem A. (Signal.) *2 times what number equals 20.*
- What number will you count by? (Signal.) *2.*
- What number will you count to? (Signal.) *20.*
- What will you make each time you count? (Signal.) *A dot.*
(Repeat until firm.)

c. Make the dots as you count by twos to 20. Get ready. (Signal.) *2 ... 4 ... 6 ... 8 ... 10 ... 12 ... 14 ... 16 ... 18 ... 20.*
- Everybody, count the dots and write the number of times on the line.
(Observe students and give feedback.)
- Everybody, 2 times what number equals 20? (Signal.) *10.*
- Say the equation. (Signal.) *2 x 10 = 20.*

from Lesson 91, Exercise 3

Teaching Note: Make sure students respond firmly to the pair of questions in step B:

"What number will you count by?"

"What number will you count to?"

A good plan is to repeat this step, even if students seem firm. If they are confused, tell them that the problem tells the number they will count to. That's 20.

Students work on these problems with reduced structure through Lesson 96. Later, they work sets of multiplication problems that have both problems that give the factor and problems with a missing factor.

Here's the set of problems they work on Lesson 101:

a.	4×6 = ▓	d.	5×8 = ▓
b.	9×▓ = 63	e.	10×▓ = 60
c.	2×▓ = 10	f.	9×4 = ▓

Textbook Lesson 101, Part 3

The work with multiplication and missing factor is connected with work on area. The displays for this work involve rectangles that show the top row, but the rest of the rectangle is covered with a patch. For some of the problems, there is a statement that tells about the number of rows. For others, there is a statement about the number of squares in the whole rectangle.

Here's part of the exercise that introduces problems that tell about the number of rows:

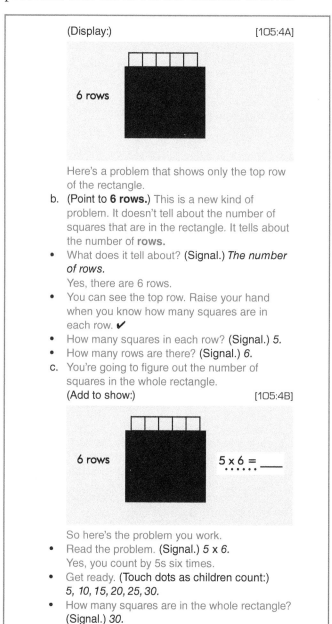

(Display:)   [105:4A]

Here's a problem that shows only the top row of the rectangle.
b. (Point to **6 rows.**) This is a new kind of problem. It doesn't tell about the number of squares that are in the rectangle. It tells about the number of **rows.**
• What does it tell about? (Signal.) *The number of rows.*
  Yes, there are 6 rows.
• You can see the top row. Raise your hand when you know how many squares are in each row. ✔
• How many squares in each row? (Signal.) *5.*
• How many rows are there? (Signal.) *6.*
c. You're going to figure out the number of squares in the whole rectangle.
  (Add to show:)   [105:4B]

So here's the problem you work.
• Read the problem. (Signal.) *5 x 6.*
  Yes, you count by 5s six times.
• Get ready. (Touch dots as children count:) *5, 10, 15, 20, 25, 30.*
• How many squares are in the whole rectangle? (Signal.) *30.*

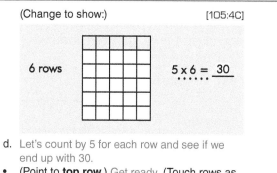

(Change to show:)   [105:4C]

6 rows $5 \times 6 = \underline{30}$

d. Let's count by 5 for each row and see if we end up with 30.
• (Point to **top row.**) Get ready. (Touch rows as students count:) *5, 10, 15, 20, 25, 30.*
• Are there 30 squares in the rectangle? (Signal.) *Yes.*

from Lesson 105, Exercise 4

On the following lesson, students work a mixed set that has problems that tell about the number of rows and problems that tell about the total number of squares.

Here's part of that exercise:

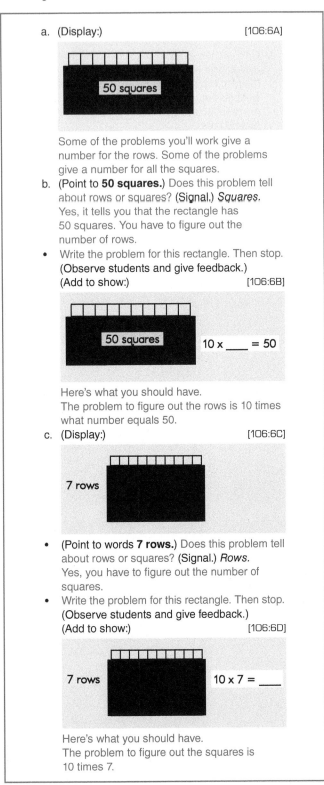

from Lesson 106, Exercise 6

The final application in the Multiplication track involves an extension of the area/row problems to pictures of objects. The problem gives either the total number of objects or the number of groups. Students translate the problem into either a multiplication problem or a missing factor problem. This work begins on Lesson 109.

Here's part of the exercise from Lesson 116:

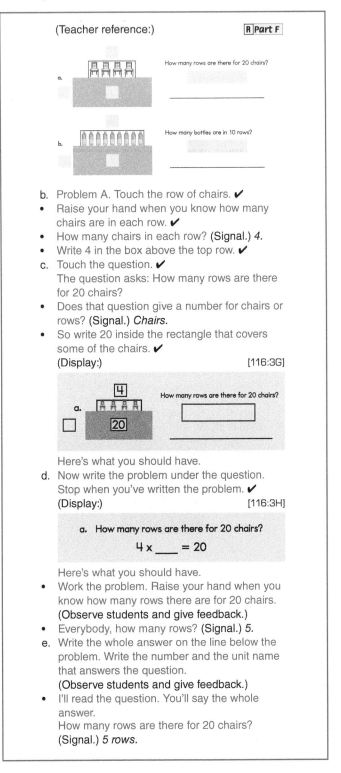

How many rows are there for 20 chairs?
(Signal.) *5 rows.*
f. Problem B. Touch the row of bottles. ✔
• Raise your hand when you know how many
  bottles are in each row. ✔
• How many bottles? (Signal.) *9.*
• So write 9 in the box above the top row. ✔
g. Touch the question. ✔
    The question asks: How many bottles are in
    10 rows?
• Does that question give a number for bottles
  or rows? (Signal.) *Rows.*
• What's the number for rows? (Signal.) *10.*
• Write the number next to the rectangle. ✔
    (Display:)   [116:3I]

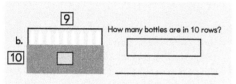

    Here's what you should have.
    There are 9 bottles in each row, and there are
    10 rows.
h. Write the problem under the question.
    Then stop. ✔
    (Display:)   [116:3J]

> **b. How many bottles are in 10 rows?**
>
> $9 \times 10 =$ ___

    Here's what you should have.
• Work the problem. Raise your hand when you
  know how many bottles are in 10 rows.
  (Observe students and give feedback.)
• Everybody, how many bottles? (Signal.) *90.*
i. Write the whole answer on the line below the
    problem. Write the number and the unit name
    that answers the question.
    (Observe students and give feedback.)
• I'll read the question. You'll say the whole
  answer.
  How many bottles are in 10 rows? (Signal.)
  90 bottles.

from Lesson 116, Exercise 3

Teaching Note: Notice that the problems look
similar to the earlier area problems. Students
write the information the problem gives in the
appropriate box in the diagram. The position
of the numbers that are given prompts whether
the problem is a multiplication or missing-
factor problem. If students have trouble, ask
them, "Does the problem give a number for
the rows or for the objects? Right, it gives a
number for all the objects. You have to figure
out the number of rows. Write that problem."

Measurement (Lessons 30–102)

This track begins on Lesson 30 and continues
intermittently through Lesson 102. Students learn
to use a ruler to measure lengths in inches and
centimeters. They learn about the size relationship
of inches and centimeters. They learn about
perimeter and area. The units for perimeter are
linear; the units for area are square.

Work with inches begins on Lesson 30. The first
activity involves marking inches on a line. The
teacher presents some examples on the board,
and then students mark inches on a line that is
positioned above a ruler.

Here's the Workbook display students start with.

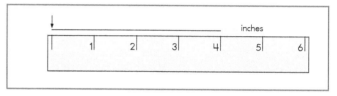

Workbook Lesson 30, Part 6

Here's how they mark the line.

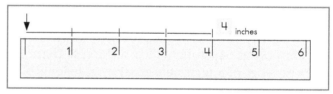

from Lesson 30, Exercise 8

Starting on Lesson 34, students learn about the
size relationship of inches and centimeters. The
rule students learn is that two inches is about five
centimeters.

Here's the Workbook material for measurement on Lesson 35:

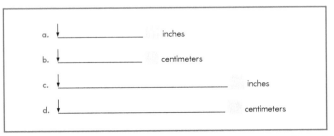

Workbook Lesson 35, Part 5

Students use rulers that show both inches and centimeters. They measure lines that are about four inches and ten centimeters.

On Lesson 39, students measure parts of a line in inches and parts in centimeters.

Here's the line from Lesson 39:

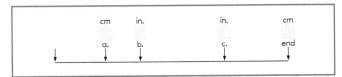

Workbook Lesson 39, Part 5

To perform the measuring, students have to flip their ruler so the correct units are shown on top. On Lesson 64, students learn the rule that opposite sides of a rectangle are the same length.

Here's the exercise from Lesson 65:

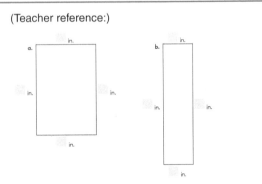

(Teacher reference:)

- Are these triangles? (Signal.) *No.*
- What are they? (Signal.) *Rectangles.*
- b. Listen: If the top side of a rectangle is 6 inches long, what other side is 6 inches long? (Signal.) *The bottom side.*
- If one up-and-down side is 4 inches long, how long is the other up-and-down side? (Signal.) *4 inches.*
- c. Touch rectangle A. ✔
- Measure each side. Write the number of inches in the boxes. (Observe students and give feedback.)
- d. Check your work.
- What is the length of the top side and bottom side? (Signal.) *2 inches.*
- What is the length of the sides that go up and down? (Signal.) *3 inches.*
- e. Touch rectangle B. ✔
- Write the length of each side in the boxes. (Observe students and give feedback.)
- f. Check your work.
- What is the length of the top side and bottom side? (Signal.) *1 inch.*
- What is the length of the sides that go up and down? (Signal.) *4 inches.*

Lesson 65, Exercise 5

Students measure each side and write the number in the box.

Students will later apply the rule about sides of a rectangle when they work perimeter problems involving rectangles.

Here's part of the perimeter exercise from Lesson 70:

- Say the problem for the up-and-down sides.
  (Signal.) *3 + 3.*
  So I write 3 plus 3.
  (Add to show:)   [70:2F]

  1 in.

  3 in. ⬚ 3 in.   1 + 1
                       3 + 3

  1 in.

k. Now we write the answer for each problem.
- (Point to **1 + 1**.) What's 1 plus 1? (Signal.) *2.*
  (Add to show:)   [70:2G]

  1 in.

  3 in. ⬚ 3 in.   1 + 1 = 2
                       3 + 3

  1 in.

- (Point to **3 + 3**.) What's 3 plus 3? (Signal.) *6.*
  (Add to show:)   [70:2H]

  1 in.

  3 in. ⬚ 3 in.   1 + 1 = 2
                       3 + 3 = 6

  1 in.

  The first answer tells about two sides. The second answer tells about the other two sides. So we add those lengths to find the length of all the sides.
- We work the problem 2 plus 6.
  Say the problem we work. (Signal.) *2 + 6.*
l. (Point to **2** and **6**.) What's 2 plus 6? (Signal.) *8.*
  (Add to show:)   [70:2I]

  1 in.

  3 in. ⬚ 3 in.   1 + 1 = 2
                       3 + 3 = 6
                              ─────
                              8 in.

  1 in.

- So the perimeter is 8 inches.
  What's the perimeter? (Signal.) *8 inches.*
m. Say the equation for the top and bottom sides.
  (Signal.) *1 + 1 = 2.*
- Say the equation for the up-and-down sides.
  (Signal.) *3 + 3 = 6.*
- Say the equation for 2 plus 6. (Signal.)
  2 + 6 = 8.
- So what's the perimeter? (Signal.) *8 inches.*

from Lesson 70, Exercise 2

Teaching Note: Students add the opposite sides (1 + 1 and 3 + 3). They write the answer to each problem and then add the totals (2 + 6). This routine has quite a few steps, so students may become confused. Remind them of the steps: "You add the top and bottom sides. Then you add the up and down sides. Then you add the answers. Once more: You add. . . ."

After students have worked with rectangles for seven lessons, they find the perimeter of triangles. For this procedure, they add the three sides.

Here's part of the exercise from Lesson 77:

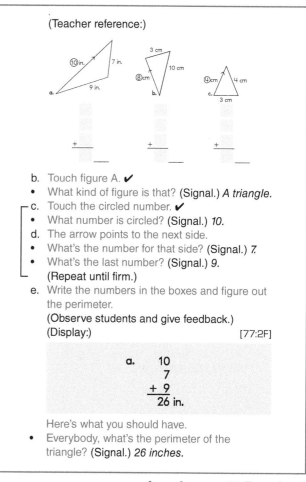

b. Touch figure A. ✔
- What kind of figure is that? (Signal.) *A triangle.*
c. Touch the circled number. ✔
- What number is circled? (Signal.) *10.*
d. The arrow points to the next side.
- What's the number for that side? (Signal.) *7.*
- What's the last number? (Signal.) *9.*
  (Repeat until firm.)
e. Write the numbers in the boxes and figure out the perimeter.
  (Observe students and give feedback.)
  (Display:)   [77:2F]

  a.   10
           7
        + 9
        ──────
         26 in.

  Here's what you should have.
- Everybody, what's the perimeter of the triangle? (Signal.) *26 inches.*

from Lesson 77, Exercise 2

This exercise follows a more structured exercise in which they say the problem for all three sides. They start with the side that has the circled number. They follow the arrow to the next number, say the addition for those numbers, then continue to the third side and add that number.

On Lesson 79, students work a mixed set of problems. Two are rectangles; two are triangles. This is the first time perimeter problems appear in the Textbook.

Here's the exercise:

(Teacher reference:)

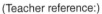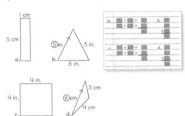

Some of these figures are rectangles, and some are triangles. You're going to figure out their perimeters.
b. Touch problem A. ✔
The picture shows how you write this problem on your lined paper.
• Write the two equations. Then stop.
(Observe students and give feedback.)
(Display:)   [79:6A]

> a. $1 + 1 = 2$
> $5 + 5 = 10$

Here's what you should have.
The top equation is $1 + 1 = 2$. The other equation is $5 + 5 = 10$.
c. Work the problem for all four sides and write the answer. Remember the unit name.
(Add to show:)   [79:6B]

> a. $1 + 1 = 2$
> $5 + 5 = \underline{10}$
> 12 cm

Here's what you should have.
• What's the perimeter of the rectangle?
(Signal.) *12 centimeters.*

d. Touch problem B. ✔
• Write the column problem. Then stop. ✔
(Display:)   [79:6C]

> b. 5
> 5
> $+ 6$

Here's what you should have.
e. Work the problem and write the answer.
Remember the unit name.
(Observe students and give feedback.)
(Add to show:)   [79:6D]

> b. 5
> 5
> $+ 6$
> $\overline{16 \text{ in.}}$

Here's what you should have.
• What's the perimeter? (Signal.) *16 inches.*
f. Touch problem C. ✔
• Write the problem and work it.
(Observe students and give feedback.)
(Display:)   [79:6E]

> c. $4 + 4 = 8$
> $4 + 4 = \underline{8}$
> 16 in.

Here's what you should have.
• What's the perimeter? (Signal.) *16 inches.*
g. Touch problem D. ✔
• Write the problem and work it.
(Observe students and give feedback.)
(Display:)   [79:6F]

> d. 6
> 3
> $+ 4$
> $\overline{13 \text{ cm}}$

Here's what you should have.
• What's the perimeter? (Signal.)
13 centimeters.

from Lesson 79, Exercise 6

AREA

In the Multiplication track, students learn the procedure for finding the area of rectangles. They find squares in a rectangle by multiplying the number of squares in each row by the number of rows. Names of square units (square inches, square centimeters, and square feet) are introduced on Lesson 96.

The term *area* is introduced on Lesson 97. Students write answers with abbreviated unit names for square inches, square centimeters, and square feet.

On Lesson 99, students work with rectangles that do not show the squares and one display is not actual size.

Here's part of the exercise:

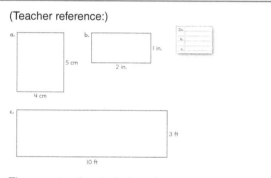

These rectangles don't show the square units. The length of two sides of each rectangle is shown. You'll figure out the area by working a times problem.

b. Rectangle A. Touch the number that tells the number of squares in each row. ✔
• What number? (Signal.) *4.*
• Start with 4 and say the times problem you work. (Signal.) *4 x 5.*
c. Rectangle B. Touch the number that tells the number of squares in each row. ✔
• What number? (Signal.) *2.*
• Start with 2 and say the times problem you work. (Signal.) *2 x 1.*
d. Rectangle C. Touch the number that tells the number of squares in each row. ✔
• What number? (Signal.) *10.*
• Start with 10 and say the times problem you work. (Signal.) *10 x 3.*
e. Find the area of rectangle A. Remember to write the unit name in the answer.
(Observe students and give feedback.)
(Display:)   [99:7A]

> a. $4 \times 5 = 20$ sq cm
>   • • • • •

Here's what you should have.
The area is 20 square centimeters.

from Lesson 99, Exercise 7

For the last problem type in this track, students find both perimeter and area of rectangles. These exercises start on Lesson 101.

Here's part of the exercise from Lesson 102:

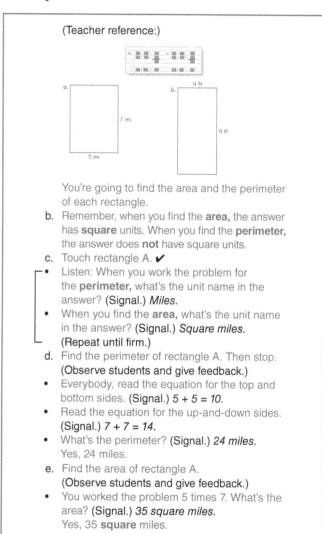

You're going to find the area and the perimeter of each rectangle.
b. Remember, when you find the **area,** the answer has **square** units. When you find the **perimeter,** the answer does **not** have square units.
c. Touch rectangle A. ✔
• Listen: When you work the problem for the **perimeter,** what's the unit name in the answer? (Signal.) *Miles.*
• When you find the **area,** what's the unit name in the answer? (Signal.) *Square miles.*
(Repeat until firm.)
d. Find the perimeter of rectangle A. Then stop. (Observe students and give feedback.)
• Everybody, read the equation for the top and bottom sides. (Signal.) *5 + 5 = 10.*
• Read the equation for the up-and-down sides. (Signal.) *7 + 7 = 14.*
• What's the perimeter? (Signal.) *24 miles.*
Yes, 24 miles.
e. Find the area of rectangle A. (Observe students and give feedback.)
• You worked the problem 5 times 7. What's the area? (Signal.) *35 square miles.*
Yes, 35 **square** miles.

from Lesson 102, Exercise 6

Inequality and Equivalence (Lessons 23–113)

The Inequality and Equivalence track starts on Lesson 23 and continues through Lesson 113. Students compare 2- and 3-digit numbers, work items involving an operation on one side, and work items involving coins. They learn to apply rules of transitivity and to work with equivalent unit relationships.

COMPARING NUMBERS AND INEQUALITY

The first activities require students to order numbers from smallest to largest or largest to smallest.

Here's part of the exercise from Lesson 24:

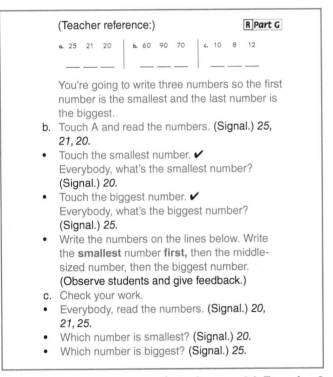

(Teacher reference:)   R Part G

a. 25 21 20 | b. 60 90 70 | c. 10 8 12
___ ___ ___   ___ ___ ___   ___ ___ ___

You're going to write three numbers so the first number is the smallest and the last number is the biggest.

b. Touch A and read the numbers. (Signal.) *25, 21, 20.*
- Touch the smallest number. ✔
  Everybody, what's the smallest number? (Signal.) *20.*
- Touch the biggest number. ✔
  Everybody, what's the biggest number? (Signal.) *25.*
- Write the numbers on the lines below. Write the **smallest** number **first,** then the middle-sized number, then the biggest number. (Observe students and give feedback.)
c. Check your work.
- Everybody, read the numbers. (Signal.) *20, 21, 25.*
- Which number is smallest? (Signal.) *20.*
- Which number is biggest? (Signal.) *25.*

from Lesson 24, Exercise 2

Teaching Note: Students should not have trouble with this exercise because they have been working with number families and are familiar with the terminology of "big number" and "small number." They also are practiced in counting sequences, counting from smaller numbers.

Starting on Lesson 27, students use the signs > and < to compare numbers (8 > 7, 7 < 8).

The rule students learn about the sign is that the bigger end of the sign is next to the bigger number. The smaller end of the sign is next to the smaller number.

Here's part of the exercise from Lesson 27:

a. (Display:)   [27:2A]

>

Here's a new sign. This sign has a smaller end and a bigger end. Tell me if I touch the smaller end or the bigger end.
- (Touch bigger.) *Bigger end.*
- (Touch smaller.) *Smaller end.*
b. (Display:)   [27:2B]

<

Here's another sign with a smaller end and a bigger end. Tell me which end I touch.
- (Touch smaller.) *Smaller end.*
- (Touch bigger.) *Bigger end.*
c. Here's a rule about these signs: The bigger end is next to the bigger number. The smaller end is next to the smaller number. (Display:)   [27:2C]

6   3

- Which number is bigger, 6 or 3? (Signal.) *6.*
  So I make the bigger end of the sign next to 6. (Add to show:)   [27:2D]

6 > 3

- Again: Tell me which number is bigger. (Signal.) *6.*
- Is the bigger end of the sign next to the bigger number? (Signal.) *Yes.*
- Is the smaller end of the sign next to the smaller number? (Signal.) *Yes.*
  So this is the right sign.
d. (Display:)   [27:2E]

8   9

- Which number is bigger, 8 or 9? (Signal.) *9.*
  So I make the bigger end of the sign next to 9. (Add to show:)   [27:2F]

8 < 9

- Again: Tell me which number is bigger. (Signal.) *9.*
- Is the bigger end of the sign next to the bigger number? (Signal.) *Yes.*
- Is the smaller end of the sign next to the smaller number? (Signal.) *Yes.*
  So this sign is right.

from Lesson 27, Exercise 2

Teaching Note: The key discrimination that students must learn early is to identify the ends of the sign—the big end and the small end. If they don't learn this relationship early, they'll have serious problems with statements that compare values. Repeat steps A and B at least one time. When you touch the small end of the sign, touch the point. When you touch the big end, touch the ends of both lines.

In step C, students learn a related discrimination: "The bigger end is next to the bigger number. The smaller end is next to the small number."

If students give weak responses when asked, "Is the bigger end of the sign next to the bigger number?", they may not see the spatial relationship between the number and the sign.

Correct weak responses by touching the bigger end of the sign and asking:

"Is this the bigger end of the sign?"

"What number is this end next to?"

"Is ___ the bigger number?"

"So the bigger end of the sign is next to the bigger number."

Practice this correction. You will probably have more than one opportunity to use it.

On the following lessons, students make signs between pairs of numbers.

On Lesson 33, students work with statements that have letters and numbers. For these, they cannot rely on these symbols to determine which is bigger. They have to rely on the sign between them.

This is the first exercise in which students say two statements for each inequality.

Here's part of the exercise from Lesson 33:

a. (Display:) [33:8A]

$$7 > 5 \quad C > B \quad R < 12$$

b. (Point to **7 > 5.**) My turn to read this statement: (Touch symbols as you say:) 7 is more than 5. I can also read it starting with the 5. (Touch symbols as you say:) 5 is less than 7.
   • (Point to **5.**) Your turn: Read the statement that starts with 5. (Touch.) *5 is less than 7.*
   • (Point to **7.**) Read the statement that starts with 7. (Touch.) *7 is more than 5.*
   • (Point to **5.**) Read the statement that starts with 5. (Touch.) *5 is less than 7.*
   (Repeat until firm.)
c. (Point to **C.**) Read the statement that starts with C. (Signal.) *C is more than B.*
   • (Point to **B.**) Read the statement that starts with B. (Signal.) *B is less than C.*
   • Which is more, B or C? (Signal.) *C.*
   (Repeat until firm.)

from Lesson 33, Exercise 8

Teaching Note: Make sure that students respond correctly on the two segments that are bracketed.

On Lesson 36, students write signs for items that have an addition problem or subtraction problem on one side. Students work these items by computing the amount on the side, crossing out the problem, writing the answer, then making the sign between the sides.

Here's part of the exercise from Lesson 36:

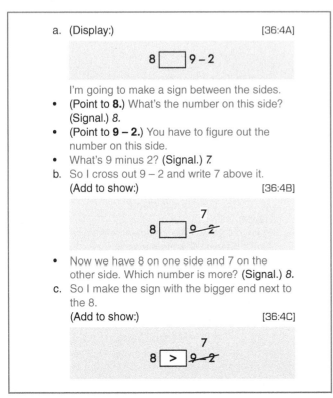

a. (Display:) [36:4A]

$$8 \boxed{} \; 9-2$$

I'm going to make a sign between the sides.
- (Point to **8**.) What's the number on this side? (Signal.) *8.*
- (Point to **9 – 2**.) You have to figure out the number on this side.
- What's 9 minus 2? (Signal.) *7*

b. So I cross out 9 – 2 and write 7 above it. (Add to show:) [36:4B]

$$8 \boxed{} \; \overset{7}{\cancel{9-2}}$$

- Now we have 8 on one side and 7 on the other side. Which number is more? (Signal.) *8.*

c. So I make the sign with the bigger end next to the 8. (Add to show:) [36:4C]

$$8 \boxed{>} \; \overset{7}{\cancel{9-2}}$$

from Lesson 36, Exercise 4

Starting on Lesson 40, students make the sign between rows of coins. Here's the set of items from Lesson 41:

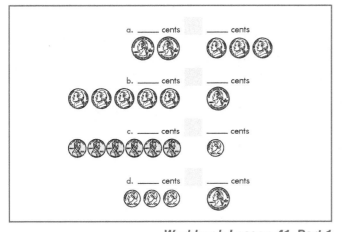

Workbook Lesson 41, Part 1

Students write the number of cents above each group, then write the sign between the pair of numbers. Note that problem B has sides that are equal.

> **Teaching Note:** If students perform well on the preceding inequality exercises, they should have no trouble with these coin problems.

On Lessons 112 and 113, students write the inequality or equal sign between two 3-digit numbers. Items include "place-value addition" statements on one side.

Here is the problem set from Lesson 113:

a. 148   100 + 80 + 4   d. 515   500 + 10 + 8

b. 600 + 70 + 8   765   e. 300 + 20 + 0   420

c. 710   700 + 0 + 9

Workbook Lesson 113, Part 4

TRANSITIVITY

Additional applications that involve the signs < and > begin on Lesson 42. These applications are labeled *transitivity*: If A is more than B and B is more than C, then A is more than C.

Here's part of the Workbook exercise from Lesson 43:

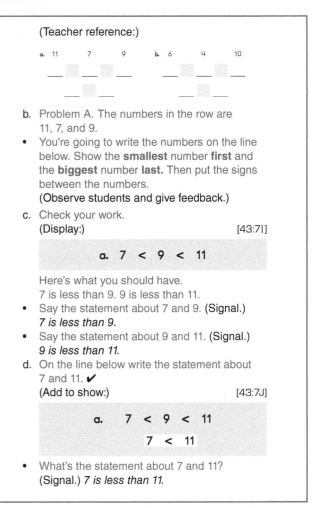

(Teacher reference:)

a. 11   7   9   b. 6   4   10

b. Problem A. The numbers in the row are 11, 7, and 9.
- You're going to write the numbers on the line below. Show the **smallest** number **first** and the **biggest** number **last**. Then put the signs between the numbers. (Observe students and give feedback.)

c. Check your work. (Display:) [43:7I]

a. 7 < 9 < 11

Here's what you should have.
7 is less than 9. 9 is less than 11.
- Say the statement about 7 and 9. (Signal.) *7 is less than 9.*
- Say the statement about 9 and 11. (Signal.) *9 is less than 11.*

d. On the line below write the statement about 7 and 11. ✔ (Add to show:) [43:7J]

a. 7 < 9 < 11
7 < 11

- What's the statement about 7 and 11? (Signal.) *7 is less than 11.*

from Lesson 43, Exercise 7

Connecting Math Concepts

Students write two statements. The first has three values arranged in order of size; the other statement has two values, the first and the last. The same sign connects all the values.

Starting on Lesson 49, students create a statement with two signs and another statement with one sign. For example, given:

$$M > V$$

$$15 > M$$

students first write a statement that relates all three values: 15 > M > V, and then write the statement without the middle value 15 > V.

The rule students follow when creating the statement with three values is that the value that appears two times is the middle value in the statement. For the example above, M is the middle value.

Here's part of the exercise from Lesson 51:

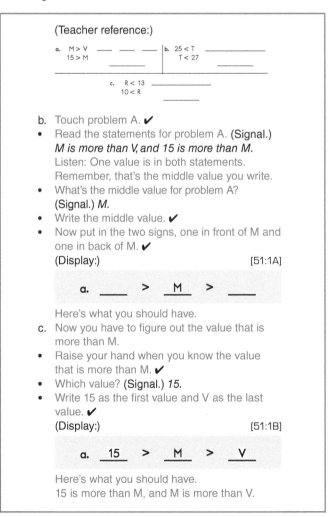

(Teacher reference:)

a. M > V ___ ___ ___   b. 25 < T _____
    15 > M   _____   T < 27
                                              _____

c. R < 13 _____
    10 < R
                     _____

b. Touch problem A. ✔
• Read the statements for problem A. (Signal.)
  M is more than V, and 15 is more than M.
  Listen: One value is in both statements.
  Remember, that's the middle value you write.
• What's the middle value for problem A?
  (Signal.) *M.*
• Write the middle value. ✔
• Now put in the two signs, one in front of M and one in back of M. ✔
  (Display:)   [51:1A]

  a. ____ > __M__ > ____

  Here's what you should have.
c. Now you have to figure out the value that is more than M.
• Raise your hand when you know the value that is more than M. ✔
• Which value? (Signal.) *15.*
• Write 15 as the first value and V as the last value. ✔
  (Display:)   [51:1B]

  a. _15_ > __M__ > __V__

  Here's what you should have.
  15 is more than M, and M is more than V.

d. My turn to say the statement that does not have the middle value: 15 is more than V. Say the statement. (Signal.) *15 is more than V.*
• Write the statement. ✔
  Remember, the second statement you write does not have the middle value, just the other two values.

from Lesson 51, Exercise 1

Teaching Note: Make sure you understand the procedure before you teach it. The goal of rewriting is to make statements that are relatively hard for students to understand much more manageable.

Instead of writing three statements in the tradition of logic

   15 > M

   M > V

   Therefore, 15 is greater than V

the students write all values in a row, with M appearing only once, and the conclusion (15 > V) becomes obvious.

For the final transitivity activity, students write the conclusion directly from a pair of statements; for example:

$$12 > P$$

$$J > 12$$

They identify the middle value (12) and write the statement for the other two values:

$$J > P$$

EQUIVALENT UNITS

Work on equivalent units starts on Lesson 88 and continues through 105. It begins with facts about equivalent units like: 12 inches = 1 foot; 24 hours = 1 day. After students learn these facts, they work problems in which they write the appropriate sign for specific relationships; for example: **14 inches ☐ 1 foot.** The missing sign is >. This work starts on Lesson 91 and continues through 105.

Here's part of the exercise from Lesson 89. It reviews the three equivalences students learned on Lesson 88 and applies the first equivalence to a ruler. In the remainder of the exercise (not shown here), students apply the rules about hours and days to clocks that show a time difference of an hour and a difference of a day.

a. (Display:)   [89:4A]

> 12 inches   = 1 foot
>
> 60 minutes = 1 hour
>
> 24 hours   = 1 day

b. (Point to **12 inches.**) Here is the rule for inches in 1 foot: 12 inches equals 1 foot.
- How many inches are in 1 foot? (Signal.) *12.*
- Say the rule about 12 inches. (Signal.) *12 inches equals 1 foot.*
- Which is smaller—an inch or a foot? (Signal.) *An inch.*

c. (Point to **60 minutes.**) Here's the rule for minutes in 1 hour: 60 minutes equals 1 hour.
- How many minutes are in 1 hour? (Signal.) *60.*
- Say the rule about 60 minutes. (Signal.) *60 minutes equals 1 hour.*
- Which is smaller—a minute or an hour? (Signal.) *A minute.*

d. (Point to **24 hours.**) Here's the rule for hours in 1 day: 24 hours equals 1 day.
- How many hours are in 1 day? (Signal.) *24.*
- Say the rule about 24 hours. (Signal.) *24 hours equals 1 day.*
- Which is smaller—an hour or a day? (Signal.) *An hour.*
  (Repeat until firm.)

e. (Change to show:)   [89:4B]

f. Let's do those without looking.
- Say the rule about inches and feet. (Signal.) *12 inches equals 1 foot.*
- Say the rule about minutes and hours. (Signal.) *60 minutes equals 1 hour.*
- Say the rule about hours and days. (Signal.) *24 hours equals 1 day.*
  (Repeat until firm.)

g. (Display:)   [89:4C]

> 0 1 2 3 4 5 6 7 8 9 10 11 12

Here's a big picture of a ruler.
- (Point to **12.**) How many inches long is this ruler? (Signal.) *12.*
- How many feet long is this ruler? (Signal.) *1.*
- Say the rule about inches and feet. (Signal.) *12 inches equals 1 foot.*

from Lesson 89, Exercise 4

Teaching Note: The question about the smaller unit in each pair is very important because it implies a relationship that students will work with later: For any given amount, there are more of the smaller units than the larger unit.

On Lesson 91, students do the first written exercise. They write the missing sign that compares two entities. Here's part of the exercise from Lesson 94:

(Teacher reference:)

a. 15 months ☐ 1 year		**d.** 58 minutes ☐ 1 hour
b. 7 days ☐ 1 week		**e.** 14 inches ☐ 1 foot
c. 23 hours ☐ 1 day		**f.** 100 cents ☐ 1 dollar

You're going to tell me the rule and the missing sign for each item.

b. Item A: 15 months and 1 year.
- Say the rule for months in a year. (Signal.) *12 months equals 1 year.*
- Is 15 months more than, less than, or equal to 1 year? (Signal.) *More than 1 year.*

c. Item B: 7 days and 1 week.
- Say the rule for days in a week. (Signal.) *7 days equals 1 week.*
- Is 7 days more than, less than, or equal to 1 week? (Signal.) *Equal to 1 week.*

d. Item C: 23 hours and 1 day.
- Say the rule for hours in a day. (Signal.) *24 hours equals 1 day.*
- Is 23 hours more than, less than, or equal to 1 day? (Signal.) *Less than 1 day.*

e. Item D: 58 minutes and 1 hour.
- Say the rule for minutes in an hour. (Signal.) *60 minutes equals 1 hour.*
- Is 58 minutes more than, less than, or equal to 1 hour? (Signal.) *Less than 1 hour.*

f. Item E: 14 inches and 1 foot.
* Say the rule for inches in a foot. (Signal.)
12 inches equals 1 foot.
* Is 14 inches more than, less than, or equal to 1 foot? (Signal.) *More than 1 foot.*
g. Item F: 100 cents and 1 dollar.
* Say the rule for cents in a dollar. (Signal.)
100 cents equals 1 dollar.
* Is 100 cents more than, less than, or equal to 1 dollar? (Signal.) *Equal to 1 dollar.*
h. Write the missing sign in each item.
(Observe students and give feedback.)
i. Check your work.
(Display:)   [94:4A]

a.	15 months	> 1 year
b.	7 days	= 1 week
c.	23 hours	< 1 day
d.	58 minutes	< 1 hour
e.	14 inches	> 1 foot
f.	100 cents	= 1 dollar

Here's what you should have for each item.
* (Point to **A.**) Everybody, read statement A. (Signal.) *15 months is more than 1 year.*
* (Point to **B.**) Read statement B. (Signal.) *7 days equals 1 week.*
* (Point to **C.**) Read statement C. (Signal.) *23 hours is less than 1 day.*
* (Point to **D.**) Read statement D. (Signal.) *58 minutes is less than 1 hour.*
* (Point to **E.**) Read statement E. (Signal.) *14 inches is more than 1 foot.*
* (Point to **F.**) Read statement F. (Signal.) *100 cents equals 1 dollar.*

from Lesson 94, Exercise 4

All rules are reviewed on Lesson 104 and 105. Here's the exercise from Lesson 105:

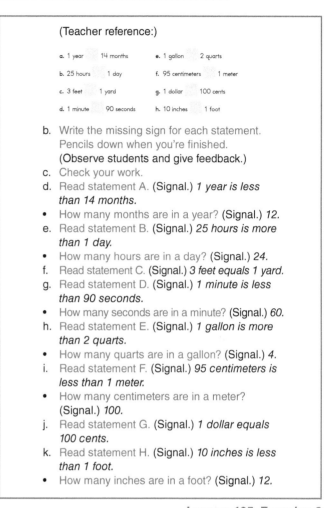

(Teacher reference:)

a. 1 year	14 months	e. 1 gallon	2 quarts
b. 25 hours	1 day	f. 95 centimeters	1 meter
c. 3 feet	1 yard	g. 1 dollar	100 cents
d. 1 minute	90 seconds	h. 10 inches	1 foot

b. Write the missing sign for each statement. Pencils down when you're finished.
(Observe students and give feedback.)
c. Check your work.
d. Read statement A. (Signal.) *1 year is less than 14 months.*
* How many months are in a year? (Signal.) *12.*
e. Read statement B. (Signal.) *25 hours is more than 1 day.*
* How many hours are in a day? (Signal.) *24.*
f. Read statement C. (Signal.) *3 feet equals 1 yard.*
g. Read statement D. (Signal.) *1 minute is less than 90 seconds.*
* How many seconds are in a minute? (Signal.) *60.*
h. Read statement E. (Signal.) *1 gallon is more than 2 quarts.*
* How many quarts are in a gallon? (Signal.) *4.*
i. Read statement F. (Signal.) *95 centimeters is less than 1 meter.*
* How many centimeters are in a meter? (Signal.) *100.*
j. Read statement G. (Signal.) *1 dollar equals 100 cents.*
k. Read statement H. (Signal.) *10 inches is less than 1 foot.*
* How many inches are in a foot? (Signal.) *12.*

Lesson 105, Exercise 3

Teaching Note: If students are not firm on some of the verbal items, repeat them before presenting step H.

As more equivalence relationships are introduced in the program, they are added to the item set. These items include 4 quarts = 1 gallon; 3 feet = 1 yard; 60 seconds = 1 minute; 100 centimeters = 1 meter.

In some of the later lessons, the items present the larger unit on the left of the sign, rather than the right.

Geometry (Lessons 43–129)

The Geometry track begins with a review on Lesson 43 and continues through Lesson 129. Students review 2-dimensional shapes, learn 3-dimensional shapes, and partition shapes into their component parts.

TWO-DIMENSIONAL SHAPES

The first shapes reviewed are rectangle, triangle, and circle.

On Lesson 46, square is introduced. The rule students learn is that all four sides of a square are the same length.

Following the introduction, students label figures in their Workbook. They write the letters **R, T,** and **C** for rectangles, triangles, and circles. For rectangles that are squares, students also write **S**.

Here's the introductory exercise from Lesson 46:

a. (Display:) [46:2A]

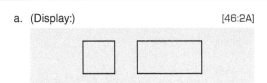

These are rectangles.
- (Point to first rectangle.) Is this a rectangle? (Signal.) *Yes.*
- (Point to second rectangle.) Is this a rectangle? (Signal.) *Yes.*
b. (Point to square.) This is a special kind of rectangle. All the sides are the same length.
- (Point to top.) This side is 5 units long.
- (Point to right side.) This side is 5 units long. And the other sides are 5 units long.
c. This rectangle is a **square.**
Say that. (Signal.) *This rectangle is a square.*
- What kind of rectangle is this? (Signal.) *A square.*
- (Point to second rectangle.) Is this a square? (Signal.) *No.*
d. Remember, all four sides of a square are the same length.

a. Find part 2 in your workbook. ✔
(Teacher reference:)

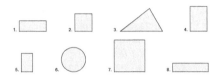

Some of these shapes are triangles; some are rectangles; and some are circles.
b. Write **R** in all the shapes that are rectangles, even the squares. Remember, the squares are rectangles.
(Observe students and give feedback.)
c. Check your work.
- Is shape 1 a rectangle? (Signal.) *Yes.*
- Is shape 2 a rectangle? (Signal.) *Yes.*
- Is shape 3 a rectangle? (Signal.) *No.*
- Is shape 4 a rectangle? (Signal.) *Yes.*
- Is shape 5 a rectangle? (Signal.) *Yes.*
- Is shape 6 a rectangle? (Signal.) *No.*
- Is shape 7 a rectangle? (Signal.) *Yes.*
- Is shape 8 a rectangle? (Signal.) *Yes.*
d. Write **T** in all the triangles and **C** in all the circles. ✔
e. Check your work.
- What's the number of the triangle? (Signal.) *3.*
- What's the number of the circle? (Signal.) *6.*
f. Listen: Some of the rectangles are squares. Remember, all sides of a square are the same length.
- Write **S** inside all the rectangles that are squares.
(Observe students and give feedback.)
g. Check your work.
- What's the number of the first square? (Signal.) *2.*
- What's the number of the other square? (Signal.) *7.*
You should have **R** and **S** inside figures 2 and 7. You shouldn't have two letters inside any other rectangles.

Lesson 46, Exercise 2

In the workbook practice, students first label all the rectangles (step B). Then students label triangles and circles (step D). Finally, students write S in each rectangle that is a square. This order insures that students won't have trouble with the idea that they are to write two letters in squares.

Students work daily exercises involving rectangle, triangle, circle, and square through Lesson 48.

THREE-DIMENSIONAL SHAPES

On Lesson 50, three-dimensional shapes are introduced—cube, pyramid, and sphere.

Here's the exercise from Lesson 51:

a. (Display:)   [51:6A]

These shapes are not flat.
b. My turn to say the names of these shapes. (Touch each shape as you say:) *Cube, pyramid, pyramid, sphere.*
•   Your turn: Say the names of these shapes. (Touch each shape as students say:) *Cube, pyramid, pyramid, sphere.*
c. Listen: I'm going to touch a shape. Each face of this shape is a triangle.
•   What's the name of the shape I'm going to touch? (Signal.) *Pyramid.*
•   (Touch **second pyramid.**) What's this shape? (Signal.) *Pyramid.*
d. Listen: I'm going to touch another shape. Each face of this shape is a square.
•   What's the name of the shape I'm going to touch? (Signal.) *Cube.*
•   (Touch **cube.**) What's this shape? (Signal.) *Cube.*
e. Listen: I'm going to touch another shape. Each face of this shape is a triangle.
•   What's the name of the shape I'm going to touch? (Signal.) *Pyramid.*
•   (Touch **first pyramid.**) What's this shape? (Signal.) *Pyramid.*
    I'm going to touch another shape that has triangles for faces.
•   (Touch **second pyramid.**) What's this shape? (Signal.) *Pyramid.*
f. Listen: I'm going to touch a shape that looks like a ball.
•   What's the name of the shape I'm going to touch? (Signal.) *Sphere.*
•   (Touch **sphere.**) What's the shape? (Signal.) *Sphere.*
    (Repeat until firm.)

Lesson 51, Exercise 6

Teaching Note: This is only the second day that students have worked with three-dimensional shapes, so it begins with a teacher model.

There are two types of pyramids: one has a triangular base, the other a rectangular base. Their common feature is that the side faces come to a point.

The script refers to "faces" of the various shapes. If students seem confused, tell them that a face is a flat side. The bottom side of a cube is the bottom face of the cube.

On Lesson 56, rectangular prisms are introduced. Students learn that the relationship between square and rectangle is parallel to their three-dimensional counterparts. All cubes are also rectangular prisms.

Here's the exercise that follows the introduction of rectangular prisms:

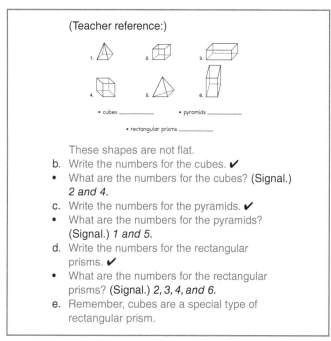

from Lesson 56, Exercise 2

Teaching Note: Students write the numbers for the appropriate shapes. For cubes, they write 2 and 4. Make sure that they also write these numbers for rectangular prisms: 2, 3, 4, and 6.

Other two-dimensional shapes are introduced later in the program, starting on Lesson 110. The shapes in this group are pentagon, hexagon, and quadrilateral.

The most diverse shape is quadrilateral, which includes all rectangles (and, therefore, all squares) and other shapes with four straight sides.

Here's the exercise from Lesson 129. This is a review of pentagon, hexagon, and quadrilateral.

a. (Display:) [129:1A]

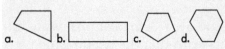

b. (Point to **A**.) What's the name of this shape? (Signal.) *Quadrilateral.*
 • Is this quadrilateral a rectangle? (Signal.) *No.*
c. (Point to **B**.) Is this shape a quadrilateral? (Signal.) *Yes.*
 • What kind of quadrilateral is this shape? (Signal.) *A rectangle.*
d. (Point to **C**.) Raise your hand when you know the name of this shape. ✔
 • What is the name of this shape? (Signal.) *Pentagon.*
 • How many sides does a pentagon have? (Signal.) *5.*
 • Say the statement about a pentagon. (Signal.) *A pentagon has 5 sides.*
e. (Point to **D**.) Raise your hand when you know the name of this shape. ✔
 • What is the name of this shape? (Signal.) *Hexagon.*
 • How many sides does a hexagon have? (Signal.) *6.*
 • Say the statement about a hexagon. (Signal.) *A hexagon has 6 sides.*
f. Once more.
 • (Point to **A**.) What's the name of this shape? (Signal.) *Quadrilateral.*
g. (Point to **B**.) Is this shape a quadrilateral? (Signal.) *Yes.*
 • What's the other name of this shape? (Signal.) *Rectangle.*
 • How many sides does it have? (Signal.) *4.*
h. (Point to **C**.) What's the name of this shape? (Signal.) *Pentagon.*
 • How many sides does it have? (Signal.) *5.*
i. (Point to **D**.) What's the name of this shape? (Signal.) *Hexagon.*
 • How many sides does it have? (Signal.) *6.*

from Lesson 129, Exercise 1

SHAPE DECOMPOSITION

On Lessons 111 through 118, students work on decomposing complex shapes into their familiar components; for example:

The component shapes are a rectangle and two triangles.

Here's part of the exercise from Lesson 111:

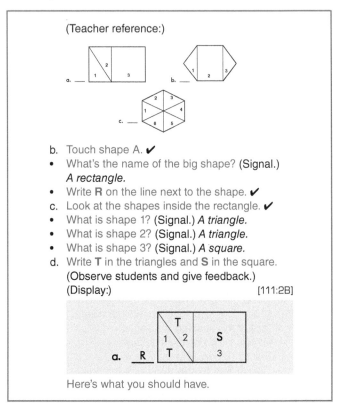

b. Touch shape A. ✔
 • What's the name of the big shape? (Signal.) *A rectangle.*
 • Write **R** on the line next to the shape. ✔
c. Look at the shapes inside the rectangle. ✔
 • What is shape 1? (Signal.) *A triangle.*
 • What is shape 2? (Signal.) *A triangle.*
 • What is shape 3? (Signal.) *A square.*
d. Write **T** in the triangles and **S** in the square. (Observe students and give feedback.) (Display:) [111:2B]

Here's what you should have.

from Lesson 111, Exercise 2

The students write the initial inside each numbered shape. They do not actually "decompose" the complex shape. They simply identify the component shapes.

Starting on Lesson 112, students show decompositions by drawing lines to create component shapes.

Here's part of the exercise from Lesson 113:

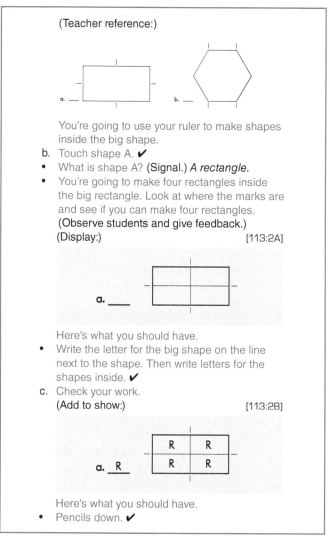

(Teacher reference:)

You're going to use your ruler to make shapes inside the big shape.

b. Touch shape A. ✔
• What is shape A? (Signal.) *A rectangle.*
• You're going to make four rectangles inside the big rectangle. Look at where the marks are and see if you can make four rectangles.
(Observe students and give feedback.)
(Display:)   [113:2A]

a. ___

Here's what you should have.
• Write the letter for the big shape on the line next to the shape. Then write letters for the shapes inside. ✔
c. Check your work.
(Add to show:)   [113:2B]

a. R

Here's what you should have.
• Pencils down. ✔

from Lesson 113, Exercise 2

Starting on Lesson 115, students work with shapes that provide options for the way they are partitioned to achieve specified decompositions.

Here's the exercise from Lesson 117:

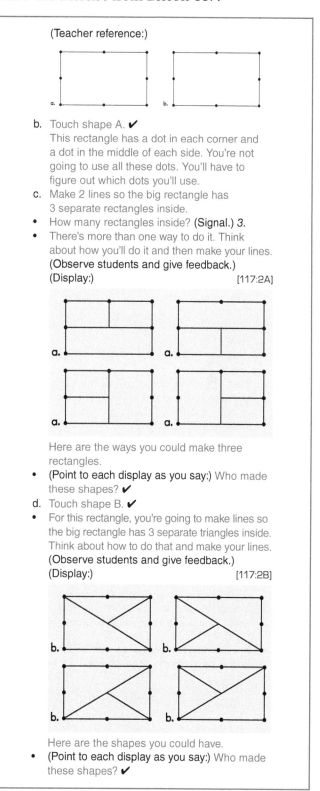

(Teacher reference:)

b. Touch shape A. ✔
This rectangle has a dot in each corner and a dot in the middle of each side. You're not going to use all these dots. You'll have to figure out which dots you'll use.
c. Make 2 lines so the big rectangle has 3 separate rectangles inside.
• How many rectangles inside? (Signal.) *3.*
• There's more than one way to do it. Think about how you'll do it and then make your lines.
(Observe students and give feedback.)
(Display:)   [117:2A]

Here are the ways you could make three rectangles.
• (Point to each display as you say:) Who made these shapes? ✔
d. Touch shape B. ✔
• For this rectangle, you're going to make lines so the big rectangle has 3 separate triangles inside. Think about how to do that and make your lines.
(Observe students and give feedback.)
(Display:)   [117:2B]

Here are the shapes you could have.
• (Point to each display as you say:) Who made these shapes? ✔

Lesson 117, Exercise 2

Students divide the rectangle into three smaller rectangles, one of which is larger than the other two. Display [117:2B] shows all four options that students could have selected.

ANGLES

Angles are introduced on Lesson 123. Students learn that lines that come to a point form an angle. They then observe that familiar shapes have the same number of angles as sides.

Here's the introductory exercise from Lesson 123:

a. (Display:)   [123:6A]

Lines that come to a point show an angle.
• Say angle. (Signal.) *Angle.*
(Touch **point.**) Here's the angle for these two lines:

b. (Display:)   [123:6B]

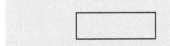

Here's a triangle.
• How many sides does a triangle have? (Signal.) *3.*

c. The triangle also has 3 angles. (Touch angles as you say:) This is one angle. This is the second angle. This is the third angle.
• A triangle has 3 angles.
Say the sentence. (Signal.) *A triangle has 3 angles.*

d. (Display:)   [123:6C]

• What shape is this? (Signal.) *A rectangle.*
• How many sides does a rectangle have? (Signal.) *4.*
• So a rectangle has 4 angles. How many angles does a rectangle have? (Signal.) *4.*

e. I'll touch the angles. You count them. Get ready. (Touch angles as students count:) *1, 2, 3, 4.*
• A rectangle has 4 angles. Say the sentence. (Signal.) *A rectangle has 4 angles.*

f. (Display:)   [123:6D]

• What shape is this? (Signal.) *A hexagon.*
• How many sides does a hexagon have? (Signal.) *6.*
• So how many angles does a hexagon have? (Signal.) *6.*

g. I'll touch the angles. You count them. Get ready. (Touch angles as students count:) *1, 2, 3, 4, 5, 6.*
• A hexagon has 6 angles. Say the sentence. (Signal.) *A hexagon has 6 angles.*

Lesson 123, Exercise 6

In later lessons, students determine whether a figure is a familiar shape based on both the number of sides and the number of angles.

Here's part of the exercise from Lesson 126:

e. (Display:)   [126:2C]

Some of these shapes are quadrilaterals.
• How many sides does a quadrilateral have? (Signal.) *4.*
Yes, a quadrilateral has 4 sides.
• What else does a quadrilateral have? (Signal.) *4 angles.*
Yes, a quadrilateral has 4 sides and 4 angles.
• Say the rule about a quadrilateral. (Signal.) *A quadrilateral has 4 sides and 4 angles.*

f. (Point to **A.**) Is this shape a quadrilateral? (Signal.) *Yes.*

g. (Point to **B.**) Is this shape a quadrilateral? (Signal.) *No.*
Why not? (Call on a student. Idea: *It doesn't have 4 sides and 4 angles; it has 5 sides and 5 angles.*)
It doesn't have 4 sides and 4 angles.
• Everybody, how many sides and angles does it have? (Signal.) *5.*

h. (Point to **C.**) Is this shape a quadrilateral? (Signal.) *Yes.*
• How do you know? (Call on a student. Idea: *It has 4 sides and 4 angles.*)

i. (Point to **D.**) Is this shape a quadrilateral? (Signal.) *No.*
• Why not? (Call on a student. Idea: *It does not have 4 angles.*)
Right. It does not have 4 angles.
• How many sides does it have? (Signal.) *4.*
• How many angles does it have? (Signal.) *3.*

from Lesson 126, Exercise 2

Teaching Note: The examples are carefully selected to ensure that students apply the rule that both conditions must be met (4 sides and 4 angles) for a figure to be a quadrilateral. Make sure students can say the rule in step E and can respond correctly to the questions in step I.

Missing Addend (Lessons 45–69)

MISSING ADDEND FROM PICTURES

In the Missing Addend track, students begin by working from picture representations to complete problems that have a missing addend. These problems start on Lesson 45 and continue through Lesson 69. Students work the first type of problem with a "count-on" strategy.

Count-on problems are familiar to continuing students. In *CMC Level B*, they worked a variety of addition problems in which the first number is given and counters are shown for the second number. Students start with the number for the first group and then count on for each item in the second group. The last number they say is the total for both groups.

In *CMC Level C* students first review the count-on strategy. Starting on Lesson 47, they apply the strategy to problems that have two groups. The first group is large and has a number above it. Students start with that number and count the objects in the second group.

Here's the exercise from Lesson 47:

(Teacher reference:)

a. 19   =   b. 35   =

This is a new kind of problem. You can see the big circles for the first number, but you can't count them. The number is written above them.

b. Touch the number for the big circles. ✔
• How many big circles are there? (Signal.) *19.*
• Count the small circles and write the number in the box. Then write the plus sign.
(Observe students and give feedback.)
c. Everybody, touch and read the problem. (Signal.) *19 + 5.*
  ⌐• Touch the number you'll get going. ✔
  │• Get it going. (Signal.) *Nineteeen.*
  │• Touch and count. (Signal.) *20, 21, 22, 23, 24.*
  └ (Repeat until firm.)
• How many circles are there in all? (Signal.) *24.*
d. Complete the equation. ✔
• Read fact A. Get ready. (Signal.) *19 + 5 = 24.*
e. Touch problem B. ✔
• Touch the number for the big boxes. ✔
• How many big boxes are there? (Signal.) *35.*
• Write the plus sign and the number for the small boxes.
(Observe students and give feedback.)
f. Everybody, touch and read problem B. (Signal.) *35 + 4.*
• Touch the number you'll get going. ✔
  ⌐• What number are you touching? (Signal.) *35.*
  │• Get it going. (Signal.) *Thirty-fiiive.*
  │• Touch and count. (Signal.) *36, 37, 38, 39.*
  └ (Repeat until firm.)
• How many boxes are there in all? (Signal.) *39.*
g. Write 39. ✔
• Read fact B. Get ready. (Signal.) *35 + 4 = 39.*

Lesson 47, Exercise 1

Teaching Note: The counting strategy is designed to preempt the most serious counting problem students have, which is to identify the first circle in the second group as 19 instead of 20. If students "get 19 going" before they count, they don't make this mistake.

When students get a number going, they say it slowly and "sing" it—nineteeen. If they have trouble, model the correct response.

"My turn to get 19 going: ninteeen."

"Your turn: Get 19 going." (Signal).

On later lessons, students work with displays that show part of a ruler. Students work addition problems of the form: 26 + __ = __. Students count to find the number they add, then work the addition problem.

Here are the problems they work on Lesson 53:

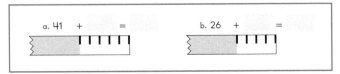

Workbook Lesson 53, Part 3

For problem A, there are 5 in the second group. Students work the problem 41 + 5. For problem B, they work the problem 26 + 4.

By Lesson 69, students work missing-addend problems by counting backwards.

Here's the first part of the exercise from Lesson 69:

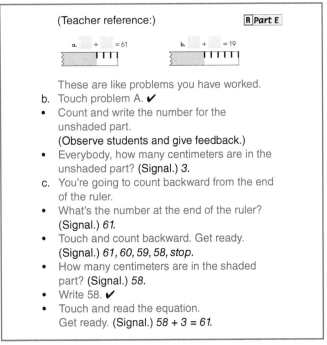

(Teacher reference:)   R Part E

a. ▨ + ▨ = 61   b. ▨ + ▨ = 19

These are like problems you have worked.
b. Touch problem A. ✔
  • Count and write the number for the unshaded part.
     (Observe students and give feedback.)
  • Everybody, how many centimeters are in the unshaded part? (Signal.) *3.*
c. You're going to count backward from the end of the ruler.
  • What's the number at the end of the ruler? (Signal.) *61.*
  • Touch and count backward. Get ready. (Signal.) *61, 60, 59, 58, stop.*
  • How many centimeters are in the shaded part? (Signal.) *58.*
  • Write 58. ✔
  • Touch and read the equation. Get ready. (Signal.) *58 + 3 = 61.*

from Lesson 69, Exercise 1

Students first figure out the number for the second group. Then they start with 61 and count backward for each unit in the second group to determine that the shaded part of the ruler is 58 units long.

MISSING ADDEND AND SUBTRAHEND ON THE NUMBER FAMILY

This track starts on Lesson 60 and continues through Lesson 76. Problems are of the form:

$$5 + \_\_ = 32$$
$$\_\_ + 11 = 40$$
$$88 - \_\_ = 24$$

The procedure students learn is to put the numbers in a number family, then solve for the missing number.

For the addition problems, the transfer is straightforward because the order of the numbers in the problem parallels the order in the number family.

$$5 + \_\_ = 32$$
$$\underline{5 \quad \longrightarrow 32}$$

The arrangement in the number family shows that you subtract to find the missing number.

For subtraction problems such as $88 - \_\_ = 24$, the order of the problem is not parallel to the order in the family. Once more, however, the solution involves subtraction.

$$\begin{array}{r} 88 \\ -24 \\ \hline \end{array}$$

Here's part of the exercise from Lesson 62. Students make number families and write the missing number in the family. Then they say the fact for the original problem. Note that they do not write the number in the original problem, only in the family.

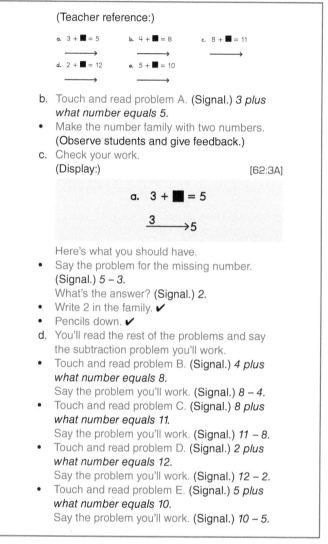

b. Touch and read problem A. (Signal.) *3 plus what number equals 5.*
• Make the number family with two numbers. (Observe students and give feedback.)
c. Check your work.
   (Display:)   [62:3A]

Here's what you should have.
• Say the problem for the missing number. (Signal.) *5 – 3.*
   What's the answer? (Signal.) *2.*
• Write 2 in the family. ✔
• Pencils down. ✔
d. You'll read the rest of the problems and say the subtraction problem you'll work.
• Touch and read problem B. (Signal.) *4 plus what number equals 8.*
   Say the problem you'll work. (Signal.) *8 – 4.*
• Touch and read problem C. (Signal.) *8 plus what number equals 11.*
   Say the problem you'll work. (Signal.) *11 – 8.*
• Touch and read problem D. (Signal.) *2 plus what number equals 12.*
   Say the problem you'll work. (Signal.) *12 – 2.*
• Touch and read problem E. (Signal.) *5 plus what number equals 10.*
   Say the problem you'll work. (Signal.) *10 – 5.*

from Lesson 62, Exercise 3

Starting on Lesson 71, students work missing-addend problems that have 2- or 3-digit numbers.

Here's part of the exercise from Lesson 72. Note that in some of these problems the first addend is missing.

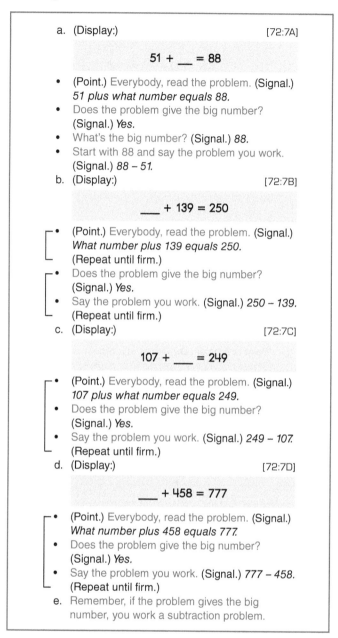

from Lesson 72, Exercise 7

Following this oral exercise, students write and work all the problems.

Subtraction problems (missing subtrahend) are introduced on Lesson 74. Students work these problems by following the rule that if the problem gives the big number, you solve it with subtraction. So, the teacher asks if the problem gives the big number. It does. So it's a subtraction problem.

Here's the first part of the exercise from Lesson 74:

a. (Display:)   [74:7A]

> a. 135 – ___ = 17
>
> b. 483 – ___ = 223
>
> c. 150 – ___ = 48
>
> d. 52 – ___ = 16

These are subtraction problems.

b. (Point to problem **A.**) Read the problem.
(Signal.) *135 minus what number equals 17.*
• Does the problem give the big number?
(Signal.) *Yes.*
• What's the big number? (Signal.) *135.*
So you figure out the missing small number by
working the problem 135 minus 17.
• Say the problem you'll work. (Signal.) *135 – 17.*
c. (Point to problem **B.**) Read the problem.
(Signal.) *483 minus what number equals 223.*
• Say the problem you'll work. (Signal.)
483 – 223.
d. (Point to problem **C.**) Read the problem.
(Signal.) *150 minus what number equals 48.*
• Say the problem you'll work. (Signal.)
150 – 48.
e. (Point to problem **D.**) Read the problem.
(Signal.) *52 minus what number equals 16.*
• Say the problem you'll work. (Signal.) *52 – 16.*

from Lesson 74, Exercise 7

Estimation (Lessons 65–127)

The Estimation track starts on Lesson 65 and
continues through Lesson 127. Students learn
procedures for rounding numbers to the nearest
ten. They use this skill to round numbers in
addition and subtraction problems. For instance,
they round the numbers in this problem: 61–18
to 60–20. Students also work the original problem
and the estimation problem, and then compare
answers to verify that the estimated answer is
close to the exact answer. Finally, they work word
problems that ask for an approximate answer.

The first activities focus on a number line from
zero to 10. Students identify the numbers that are
closer to zero or closer to 10.

Here's the exercise from Lesson 66:

a. (Display:)   [66:5A]

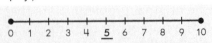

Last time, you learned three numbers that are
close to zero and three numbers that are close
to 10.
• Say the three numbers that are close to zero.
(Signal.) *1, 2, 3.*
• Say the three numbers that are close to 10.
(Signal.) *7, 8, 9.*
• (Repeat until firm.)
b. (Point to **5.**) 5 is not closer to zero or 10. It's
right in the middle.
• Is 5 closer to zero? (Signal.) *No.*
• Is 5 closer to 10? (Signal.) *No.*
c. (Point to **3.**) What is 3 closer to? (Signal.) *Zero.*
• (Point to **4.**) What is 4 closer to? (Signal.) *Zero.*
Yes, the numbers that are closer to zero are 1,
2, 3, 4.
d. Say the numbers that are closer to zero.
(Signal.) *1, 2, 3, 4.*
• Say the number that is right in the middle.
(Signal.) *5.*
• Say the numbers that are closer to 10. (Signal.)
6, 7, 8, 9.
• (Repeat until firm.)
e. I'll say a number. You'll tell me if it's closer to
zero or 10.
• 8. What's it closer to? (Signal.) *10.*
• 6. What's it closer to? (Signal.) *10.*
• 4. What's it closer to? (Signal.) *Zero.*
• 1. What's it closer to? (Signal.) *Zero.*
• 3. What's it closer to? (Signal.) *Zero.*
• 7. What's it closer to? (Signal.) *10.*
• (Repeat until firm.)
f. Listen: Say the numbers that are closer to
zero. (Signal.) *1, 2, 3, 4.*
• Say the number that is right in the middle.
(Signal.) *5.*
• Say the numbers that are closer to 10.
(Signal.) *6, 7, 8, 9.*
• (Repeat until firm.)

Lesson 66, Exercise 5

Teaching Note: Make sure students are
very firm on step F. They will use a parallel
relationship for rounding numbers from 10
through 100.

Here's part of the exercise from Lesson 74. In it, students identify whether specific numbers are closer to 50 or 60. They also indicate which number is in the middle.

a. Start with 50 and count to 60. (Signal.) *50, 51, 52, 53, 54, 55, 56, 57, 58, 59, 60.*
• Say the numbers that are closer to 50. (Signal.) *51, 52, 53, 54.*
• Say the numbers that are closer to 60. (Signal.) *56, 57, 58, 59.*
b. I'll say numbers that are between 50 and 60. You'll tell me if they are closer to 50 or closer to 60.
• Listen: 54. Is 54 closer to 50 or 60? (Signal.) *50.*
• Listen: 56. Is 56 closer to 50 or 60? (Signal.) *60.*
• What number is right in the middle? (Signal.) *55.*
c. Start with 70 and count to 80. (Signal.) *70, 71, 72, 73, 74, 75, 76, 77, 78, 79, 80.*
• Say the numbers that are closer to 70. (Signal.) *71, 72, 73, 74.*
• Listen: 77. Is 77 closer to 70 or 80? (Signal.) *80.*
• Listen: 74. Is 74 closer to 70 or 80? (Signal.) *70.*

from Lesson 74, Exercise 2

Following the oral activity, students identify the closer tens number for specific numbers. Here are the Workbook items for Lesson 74:

a. 27	20 30		e. 12	10 20
b. 34	30 40		f. 63	60 70
c. 88	80 90		g. 18	10 20
d. 26	20 30			

Workbook Lesson 74, Part 1

On the following lessons, the structure is faded. By Lesson 79, students write the tens number that is closer to the number shown. Here are the items:

a. 37	b. 42	c. 74
d. 58	e. 13	f. 86

Textbook Lesson 79, Part 2

On Lesson 80, students work their first estimation problems. Here's the exercise from Lesson 82:

a. Open your workbook to Lesson 82 and find part 1. ✔
(Teacher reference:)   ⓇPart H

You're going to write estimation problems.
b. Problem A: 36 minus 12.
• What tens number is closer to 36? (Signal.) *40.*
• What tens number is closer to 12? (Signal.) *10.*
• Write the estimation problem. Then stop. (Observe students and give feedback.)
• Everybody, read the estimation problem. (Signal.) *40 – 10.*
• What's 40 minus 10? (Signal.) *30.*
c. Problem B: 81 minus 48.
• Write the estimation problem. Then stop. (Observe students and give feedback.)
• Everybody, read the estimation problem. (Signal.) *80 – 50.*
• Raise your hand if you got it right. ✔ (Praise students who raise their hand.)
• What's 80 minus 50? (Signal.) *30.*
d. Problem C: 17 plus 41.
• Write the estimation problem and the answer. (Observe students and give feedback.)
• Everybody, read the estimation fact. (Signal.) *20 + 40 = 60.*
• Work the regular problem and see how close the answer is to 60. (Observe students and give feedback.)
• What's the answer to the regular problem? (Signal.) *58.*
That's pretty close to 60.

Lesson 82, Exercise 2

Teaching Note: You structure the work for problems A and B by asking students to identify the rounded numbers. For problem C, you present the directions, "Write the estimation problem and the answer." Note that this change in structure has a purpose. It puts more responsibility on the students to take the steps to solve the problem without teacher support.

Note also, that after students check their work, they work the original problem to confirm that the answer is close to the answer of the estimation problem.

Estimation problems first appear in Independent Work on Lesson 84.

On Lesson 91, estimation word problems are introduced. Here's the first part of the exercise from Lesson 91:

a. You've worked problems that ask how many there are. Some problems have the word **about.** They ask **about** how many there are. To work those problems, you use estimation.
b. (Display:)   [91:2A]

> There are 37 students in one classroom and
> 24 students in another classroom.
> About how many students are in both
> classrooms?

I'll read the problem: There are 37 students in one classroom and 24 students in another classroom. **About** how many students are in both classrooms?
c. This is an addition problem.
• Say the regular problem you would work with the numbers 37 and 24. (Signal.) *37 + 24.*
d. The answer to that problem tells the right number of students. The estimation problem tells a number that is **close to** the right number.
• Say the estimation problem for 37 plus 24. (Pause.) Get ready. (Signal.) *40 + 20.*
• What's the answer? (Signal.) *60.*
(Repeat until firm.)
• So **about** how many students are in both classrooms? (Signal.) *60.*

from Lesson 91, Exercise 2

Teaching Note: The word problems that follow in the exercise are of the same form as the problem presented in step B. Make sure they are firm on this problem. If they are, they will tend to perform well on saying the estimation problem for the word problems that follow.

On Lessons 93 and 94, students work problems without first saying the estimation problem. Here is the exercise from Lesson 94:

(Teacher reference:)

a. Hillary had 77 cards. Then she lost 18 cards. About how many cards did she still have?	**b.** Bill got 22 dollars last week. Then Bill got 26 dollars this week. About how many dollars did he have altogether?
_____	_____

b. I'll read both problems. Then you'll write the estimation problems and the answers.
• Problem A. Touch the words as I read them. Hillary had 77 cards. Then she lost 18 cards. About how many cards did she still have?
• Problem B. Touch the words. Bill got 22 dollars last week. Then Bill got 26 dollars this week. About how many dollars did he have altogether?
c. Work both estimation problems. Remember the unit name in the answer. Pencils down when you're finished.
(Observe students and give feedback.)
d. Check your work.
e. Say the estimation problem for A. (Signal.) *80 – 20.*
• What's the whole answer? (Signal.) *60 cards.* Yes, Hillary still had about 60 cards.
f. Say the estimation problem for B. (Signal.) *20 + 30.*
• What's the whole answer? (Signal.) *50 dollars.* So Bill had about 50 dollars altogether.

Lesson 94, Exercise 1

Teaching Note: Students do not make number families for the estimation problems because the problems are fairly simple. If students get stuck on whether they add or subtract to work a problem, tell them to make a number family and then work the problem.

Students continue to work estimation column problems and word problems through the end of the program. For estimation applied to measurement, see pages 125–133 in the **Common Core State Standards for Mathematics** section.

Telling Time (Lessons 79–123)

Exercises involving time start on Lesson 79 and continue through Lesson 123. These exercises teach telling time on analog clocks to the minute and working problems that involve a starting time and an ending time. Students learn procedures for solving problems that have times on either side of noon or midnight.

The first exercises involve telling time. Students receive information about the hands, the direction the hands move on the clock, and how to calculate minutes.

Here's part of the exercise that teaches the convention of starting at the top of the clock and counting to the hour hand:

a. (Display:)   [81:3A]

    These are clocks.
b. What's the number at the top of a clock?
    (Signal.) *12.*
• What's the next number? (Signal.) *1.*
• What's the next number? (Signal.) *2.*
• Is the hour hand the long hand or the short hand? (Signal.) *The short hand.*
c. (Point to clock **A.**) Look at the hour hand.
• I'll start at the top and move to the hour hand. Say the numbers and tell me when to stop. (Move slowly as students count:) *1, 2, 3, 4, 5, 6, 7, 8, stop.*
• What's the last number you said? (Signal.) *8.*
• So what does the hour hand show? (Signal.) *8.*

from Lesson 81, Exercise 3

On Lesson 83, students learn to count by 5 for the minute hand and learn the conventions of writing clock time.

Here's part of the exercise from Lesson 84:

a. (Display:)   [84:5A]

    (Point to clock.) I'll write the time for this clock.
• Raise your hand when you know the number for the hours. ✔
• What's the number for the hours? (Signal.) *4.*
    (Add to show:)   [84:5B]

    I made two dots after the number for the hours.
b. Now we'll figure out the minutes.
• How do we do that? (Call on a student. Idea: *Count by fives to the minute hand.*) Yes, count by fives to the minute hand.
• (Touch **12.**) Get ready. (Touch numbers as students count:) *5, 10, 15, 20, 25, 30, 35, 40.*
• What's the number for minutes? (Signal.) *40.* (Add to show:)   [84:5C]

                                 4:40

    My turn to read the time: 4 forty.
• Read the time for this clock. (Signal.) *4 forty.*

from Lesson 84, Exercise 5

In the following lessons, students practice writing clock time. For all these examples, the minute hand points to a number, so the time ends in 5 or zero.

Students next learn to read times that show o'clock times, with the minute hand at 12. On Lesson 91, students learn to read times in which the minute hand does not point to a number. For these exercises, students count by 5 to the number just before the minute hand and then count by ones to the minute hand.

---

a. (Display:)   [93:6A]

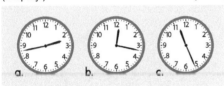

None of these minute hands points to a number. They point to a line after a number. To figure out the minutes, I touch the number just before the minute hand and figure out the minutes for that number.

b. (Point to clock **A.**) What's the number just before the minute hand? (Signal.) *8.*
So we count by 5 to that number. Then we add the minutes to the minute hand.

• (Point to **12.**) Count by fives to the number just before the minute hand. Get ready. (**Touch numbers as students count:**) *5, 10, 15, 20, 25, 30, 35, 40.*

• Now we count for each line. (**Touch lines and count:**) *41, 42, 43.*

• How many minutes? (Signal.) *43.*
Yes, the time is 2:43.

• What's the time? (Signal.) *2:43.*

c. (Point to clock **B.**) What's the number just before the minute hand? (Signal.) *3.*
So we count by fives to that number.

• (Point to **12.**) Get ready. (**Touch numbers as students count:**) *5, 10, 15.*

from Lesson 93, Exercise 6

Starting on Lesson 93, the clocks appear in the textbook, so the students must touch and count to figure out the number of minutes. Here's part of the exercise from Lesson 95:

(Teacher reference:)

You're going to figure out the minutes for each clock.

b. Clock A: Touch the number just before the minute hand. ✔

• Everybody, what number are you touching? (Signal.) *11.*

• Count by fives to that number. Get ready. (Signal.) *5, 10, 15, 20, 25, 30, 35, 40, 45, 50, 55.*

• Now count for each line to the minute hand. Get ready. (Signal.) *56, 57.*

• How many minutes does this clock show? (Signal.) *57.*

c. Clock B: Touch the number just before the minute hand. ✔

• Everybody, what number are you touching? (Signal.) *3.*

• Count by fives to that number. Get ready. (Signal.) *5, 10, 15.*

• Now count for each line to the minute hand. Get ready. (Signal.) *16, 17, 18.*

• How many minutes does this clock show? (Signal.) *18.*

from Lesson 95, Exercise 6

Students also learn facts about AM and PM. On Lesson 119, they learn that there are 24 hours in a day and that the day is divided into AM hours and PM hours. So the hour hand goes around the clock two times each day.

Here's part of the introduction of AM and PM:

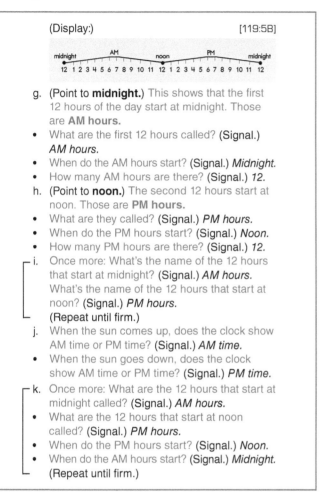

g. (Point to **midnight**.) This shows that the first 12 hours of the day start at midnight. Those are **AM hours.**
- What are the first 12 hours called? (Signal.) *AM hours.*
- When do the AM hours start? (Signal.) *Midnight.*
- How many AM hours are there? (Signal.) *12.*

h. (Point to **noon**.) The second 12 hours start at noon. Those are **PM hours.**
- What are they called? (Signal.) *PM hours.*
- When do the PM hours start? (Signal.) *Noon.*
- How many PM hours are there? (Signal.) *12.*

i. Once more: What's the name of the 12 hours that start at midnight? (Signal.) *AM hours.* What's the name of the 12 hours that start at noon? (Signal.) *PM hours.* (Repeat until firm.)

j. When the sun comes up, does the clock show AM time or PM time? (Signal.) *AM time.*
- When the sun goes down, does the clock show AM time or PM time? (Signal.) *PM time.*

k. Once more: What are the 12 hours that start at midnight called? (Signal.) *AM hours.*
- What are the 12 hours that start at noon called? (Signal.) *PM hours.*
- When do the PM hours start? (Signal.) *Noon.*
- When do the AM hours start? (Signal.) *Midnight.* (Repeat until firm.)

from Lesson 119, Exercise 5

In the following lessons, students work problems that present two times. Each item involves two clocks with different times. Students figure out the number of hours from one time to the other. If the time goes from AM to PM or PM to AM, they figure the number of hours from the starting time to 12. Then they add the ending time to that number.

Here's part of the exercise from Lesson 122:

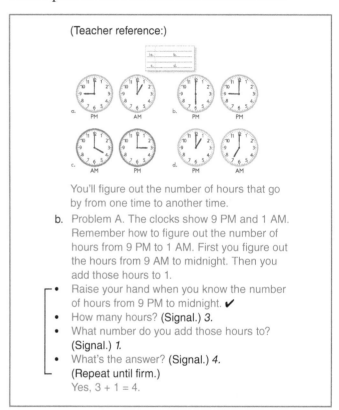

You'll figure out the number of hours that go by from one time to another time.

b. Problem A. The clocks show 9 PM and 1 AM. Remember how to figure out the number of hours from 9 PM to 1 AM. First you figure out the hours from 9 AM to midnight. Then you add those hours to 1.
- Raise your hand when you know the number of hours from 9 PM to midnight. ✔
- How many hours? (Signal.) *3.*
- What number do you add those hours to? (Signal.) *1.*
- What's the answer? (Signal.) *4.* (Repeat until firm.) Yes, 3 + 1 = 4.

from Lesson 122, Exercise 7

Data Representation and Interpretation (Lessons 100–127)

The Data Representation and Interpretation track starts on Lesson 100 and continues through Lesson 127.

Students first work with data tables that have headings for rows and columns. Students identify the two headings for items shown in the table, and follow directions for locating specific items in the table. Starting on Lesson 116, students work with picture graphs and bar graphs.

DATA TABLES

On Lesson 100, students identify columns and rows. Columns go up and down. Rows go from side to side. In the early exercises, students follow directions about crossing out or circling specific numbers.

Here's part of the exercise from Lesson 101. Students have already circled the smallest number in the moon row.

c. I'll read item B: Circle the largest number in the star row.
•   What are you going to do with the largest number in the star row? (Signal.) *Circle it.*
•   Find the star row and circle the largest number. (Observe students and give feedback.) (Add to show:)   [101:1N]

Here's what you should have.
•   The numbers are 12, 17, and 10. Which number did you circle? (Signal.) *17.*
d. I'll read item C: Cross out the largest number in column B.
•   What are you going to do with the largest number in column B? (Signal.) *Cross it out.*
•   Find column B and cross out the largest number. (Observe students and give feedback.) (Add to show:)   [101:1O]

Here's what you should have.
•   The numbers are 17, 6, and 10. Which number did you cross out? (Signal.) *17.*

from Lesson 101, Exercise 1

The next skill that students learn is to identify the "coordinates" for specific items in the table. They do this by identifying the column heading and the row heading for the item shown.

Here's part of the exercise from Lesson 104:

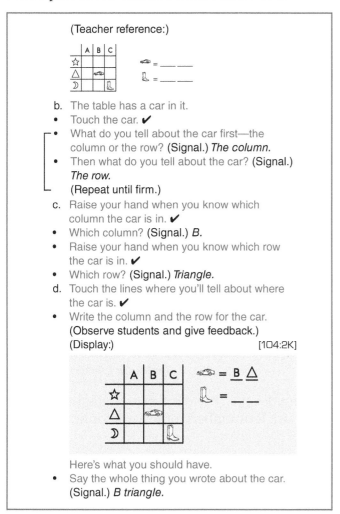

(Teacher reference:)

b. The table has a car in it.
• Touch the car. ✔
• What do you tell about the car first—the column or the row? (Signal.) *The column.*
• Then what do you tell about the car? (Signal.) *The row.*
  (Repeat until firm.)
c. Raise your hand when you know which column the car is in. ✔
• Which column? (Signal.) *B.*
• Raise your hand when you know which row the car is in. ✔
• Which row? (Signal.) *Triangle.*
d. Touch the lines where you'll tell about where the car is. ✔
• Write the column and the row for the car. (Observe students and give feedback.)
  (Display:)   [104:2K]

Here's what you should have.
• Say the whole thing you wrote about the car. (Signal.) *B triangle.*

from Lesson 104, Exercise 2

Here's part of the exercise from Lesson 106:

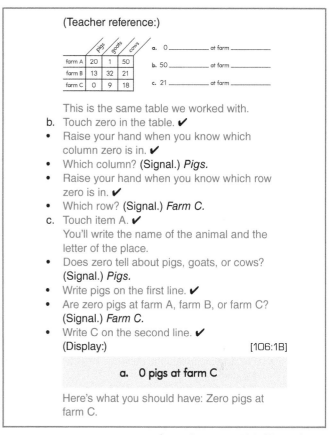

(Teacher reference:)

This is the same table we worked with.
b. Touch zero in the table. ✔
• Raise your hand when you know which column zero is in. ✔
• Which column? (Signal.) *Pigs.*
• Raise your hand when you know which row zero is in. ✔
• Which row? (Signal.) *Farm C.*
c. Touch item A. ✔
  You'll write the name of the animal and the letter of the place.
• Does zero tell about pigs, goats, or cows? (Signal.) *Pigs.*
• Write pigs on the first line. ✔
• Are zero pigs at farm A, farm B, or farm C? (Signal.) *Farm C.*
• Write C on the second line. ✔
  (Display:)   [106:1B]

a. 0 pigs at farm C

Here's what you should have: Zero pigs at farm C.

from Lesson 106, Exercise 1

Starting on Lesson 106, students work with phrases of the form:

21 _____ at farm _____.

The phrase gives the number in the table. Students complete the statement by writing the column heading and completing the row heading:

21 <u>cows</u> at farm <u>B</u>.

On Lesson 109, students identify the number in the table for phrases of the form "buses at the mall." Here's part of the exercise from Lesson 109:

a. (Display:) [109:4A]

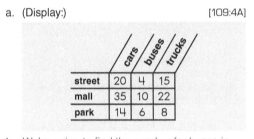

b. We're going to find the number for buses in the street.
• What's the vehicle? (Signal.) *Buses.*
• Where are the buses? (Signal.) *In the street.*
(Repeat until firm.)

c. To find buses in the street, I go to the headings buses and street. Then I go to the number.
Watch. (Touch **buses** and **street**.) (Move to **4**.)
• What's the number for buses in the street? (Signal.) *4.*

d. New problem.
• We're going to find the number for trucks at the mall.
• What's the vehicle? (Signal.) *Trucks.*
• Where are the trucks? (Signal.) *At the mall.*
(Repeat until firm.)

e. I'll touch both headings. You tell me the number.
(Touch **trucks** and **mall**.)
• What's the number for trucks at the mall? (Signal.) *22.*
(Move to **22**.) Yes, 22.

from Lesson 109, Exercise 4

Teaching Note: Your demonstration is very important because in the exercise that follows this one, students find the numbers for phrases that tell about vehicles in different places. In step C, touch both names and then move slowly to the number. Note that some students will have problems doing this. They will tend to move both hands in the same direction, rather than one going down and the other across.

In step E, students are to do it visually. If they have trouble identifying the number, move your fingers slowly toward the number. A good plan is to repeat all the items and see if students are able to identify the number as you point to the headings.

Following this demonstration, students find numbers for a table in their Textbook. They follow the procedure of touching both headings and moving to the intersection of the column and row.

The final exercise type starts on Lesson 112. This type presents questions, not phrases. All the questions in 112 are answered by a column heading or a row heading.

Here are the items:

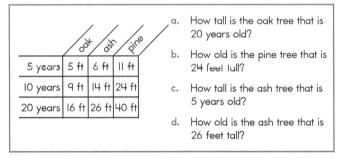

a. How tall is the oak tree that is 20 years old?
b. How old is the pine tree that is 24 feet tall?
c. How tall is the ash tree that is 5 years old?
d. How old is the ash tree that is 26 feet tall?

Textbook Lesson 112, Part 1

On later lessons, there is a mix of question types. Some ask for numbers; others ask for heading names.

PICTURE GRAPHS AND BAR GRAPHS

Work with picture graphs starts on Lesson 116. Here's an item from Lesson 116:

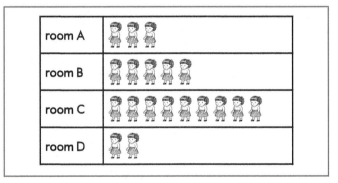

from Lesson 116, Exercise 4

Students identify the row headings and count the students in each room. Then they answer questions about which rooms have the most students and the fewest number of students.

Starting on Lesson 117, students write the numbers for the rows and then answer questions.

Bar graphs are introduced on Lesson 119. Here's a graph from Lesson 119.

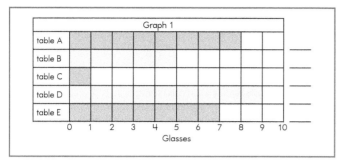

Workbook Lesson 119, Part 2

The wording the teacher uses buttresses against the most common mistakes that students make. The teacher asks two questions:

- Does any bar end at 3?
- So does any table have 3 glasses?

This question pair ensures that students tend to the total number in the rows.

On later lessons, students answer questions about *most, fewest,* and other comparisons. They also create bar graphs.

Here's the framework they start with on Lesson 123:

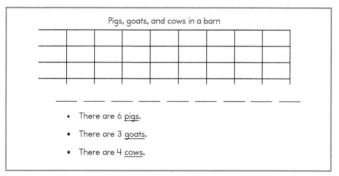

Workbook Lesson 123, Part 4

The students write the names. They write the numbers zero through 8. They shade in the bars so the bars correspond to the information provided by the statements under the graph.

LINE PLOTS

Students make line plots by measuring lines of different lengths. They first write the length at the end of each line. Students then graph the results by showing the number of lines that are different lengths.

Here's a set of 6 lines:

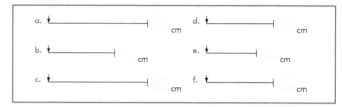

Workbook Lesson 124, Part 3

Here's a line plot that shows that data. It shows that one line is 3 cm long, three lines are 4 cm long, and two lines are 6 cm long.

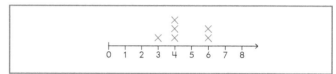

Workbook Lesson 124, Part B Answer Key

Here's a structured exercise from Lesson 125:

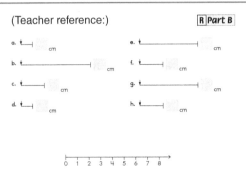

(Teacher reference:)   R Part B

* Measure each line in centimeters and write the number in the box at the end of the line. (Observe students and give feedback.)
b. Check your work. Tell me the length of each line.
* Line A. (Signal.) *1 centimeter.*
* Line B. (Signal.) *6 centimeters.*
* Line C. (Signal.) *2 centimeters.*
* Line D. (Signal.) *1 centimeter.*
* Line E. (Signal.) *5 centimeters.*
* Line F. (Signal.) *2 centimeters.*
* Line G. (Signal.) *5 centimeters.*
* Line H. (Signal.) *2 centimeters.*
c. You're going to make a line plot.
* Touch the centimeter number line in your workbook. ✔
  Remember, you make Xs above the line.
d. Touch **1** on the number line. ✔
* Are any lines 1 centimeter long? (Signal.) *Yes.*
* How many lines are 1 centimeter long? (Signal.) *2.*
* So how many Xs do you make above 1 on the number line? (Signal.) *2.*
* Make 2 **Xs** above the 1. (Observe students and give feedback.)
e. Are any lines 2 centimeters long? (Signal.) *Yes.*
* Make an **X** above 2 for each line that is 2 centimeters long. ✔
* How many Xs did you make above the 2? (Signal.) *3.*
f. Are any lines 3 centimeters long? (Signal.) *No.*
* So what do you write above the 3? (Signal.) *Nothing.*
g. You're going to check the rest of the numbers. If there are any lines that are 4 centimeters long, make an X for each line above the 4. If there are any lines that are 5 centimeters long, make an X for each line above the 5.
* Make Xs for the rest of the numbers that are on the number line. (Observe students and give feedback.)

h. Check your work. (Add to show:)   [125:4A]

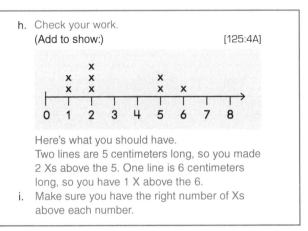

Here's what you should have.
Two lines are 5 centimeters long, so you made 2 Xs above the 5. One line is 6 centimeters long, so you have 1 X above the 6.
i. Make sure you have the right number of Xs above each number.

Lesson 125, Exercise 4

Fractions (Lessons 121–126)

Students are introduced to fractions on Lesson 121. They learn that if a figure is divided into equal parts, all the parts have a fractional name, like fourths or fifths. Students also learn that the fractional name tells how many equal parts are in the figure. If the figure has fifths, it has 5 equal parts. If the figure has thirds, it has 3 equal parts.

Here's the introductory exercise from Lesson 121:

a. (Display:) [121:5A]

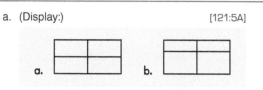

b. (Point to **A.**) Here's a rectangle.
• Raise your hand when you know how many parts are inside this rectangle. ✔
• How many parts? (Signal.) *4.*
• Listen: Are all the parts the same size? (Signal.) *Yes.*
c. (Point to **B.**) Raise your hand when you know how many parts are inside this rectangle. ✔
• How many parts? (Signal.) *4.*
• Are all the parts the same size? (Signal.) *No.* So we can't work with this rectangle.
d. (Point to **A.**) The parts are the same size in this rectangle. There are 4 parts, so each part is a fourth.
• What is each part? (Signal.) *A fourth.*
e. My turn to count the fourths. (Touch and count:) 1 fourth, 2 fourths, 3 fourths, 4 fourths.
• Your turn: Count the fourths. (Touch each part as students count:) *1 fourth, 2 fourths, 3 fourths, 4 fourths.*
• How many fourths? (Signal.) *4.*
• (Point to **B.**) These parts are not fourths because they are not the same size.

f. (Change to show:) [121:5B]

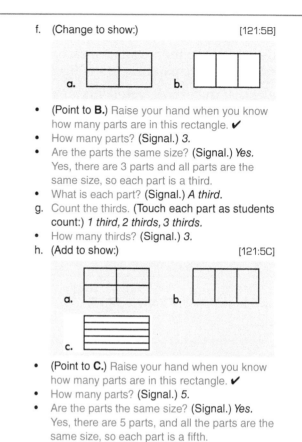

• (Point to **B.**) Raise your hand when you know how many parts are in this rectangle. ✔
• How many parts? (Signal.) *3.*
• Are the parts the same size? (Signal.) *Yes.* Yes, there are 3 parts and all parts are the same size, so each part is a third.
• What is each part? (Signal.) *A third.*
g. Count the thirds. (Touch each part as students count:) *1 third, 2 thirds, 3 thirds.*
• How many thirds? (Signal.) *3.*
h. (Add to show:) [121:5C]

• (Point to **C.**) Raise your hand when you know how many parts are in this rectangle. ✔
• How many parts? (Signal.) *5.*
• Are the parts the same size? (Signal.) *Yes.* Yes, there are 5 parts, and all the parts are the same size, so each part is a fifth.
• What is each part? (Signal.) *A fifth.*
i. Raise your hand when you know how many fifths are in this rectangle. ✔
• How many fifths? (Signal.) *5.*
j. (Point to **B.**) What is each part of this rectangle called? (Signal.) *A third.* How many thirds are in this rectangle? (Signal.) *3.*
• (Point to **A.**) What is each part of this rectangle called? (Signal.) *A fourth.* How many fourths are in this rectangle? (Signal.) *4.*
• (Point to **C.**) What is each part of this rectangle called? (Signal.) *A fifth.*
• How many fifths are in this rectangle? (Signal.) *5.*

Lesson 121, Exercise 5

Steps A through C focus on the fact that the parts must be the same size. In steps D through I, students determine the name of each part and how many of those parts are in the figure.

At the end of the exercise, students are expected to respond correctly to this pair of questions:

- What is each part of this rectangle called?
- How many ____s are in this rectangle?

Students respond to the same two questions on the following lessons.

Here are the Workbook items that students work on Lesson 125:

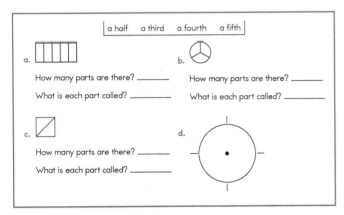

Workbook Lesson 125, Part 4

For item D, students partition the circle, first into halves, then into fourths. They say the statement for each picture:

This circle has 2 halves.
This circle has 4 fourths.

CMC Level C and Common Core State Standards for Mathematics

According to the Common Core State Standards for Mathematics:

In Grade 2, instructional time should focus on four critical areas: (1) extending understanding of base-ten notation; (2) building fluency with addition and subtraction; (3) using standard units of measure; and (4) describing and analyzing shapes.

CMC Level C meets all the Level C standards. A comprehensive listing of the standards and where they are met in the program appears on pages 490–499 of Presentation Book 1, and pages 404–412 of Presentation Book 2. Note that most of the Common Core State Standards are covered in the major tracks discussed on pages 32–124. Parts of the standards that have not already been addressed are discussed in this section. Note that text of the Common Core State Standards below has been abridged to highlight the parts of the standards that are discussed here. For a correlation to all of the standards with full text, see back of Presentation Books 1 and 2.

Operations and Algebraic Thinking (OA)

Common Core State Standards

2.OA 3:   Determine whether a group of objects (up to 20) has an odd or even number of members, e.g., by pairing objects or counting them by 2s . . .

Work with odd and even numbers begins on Lesson 124. Students learn that the numbers for counting by 2s are even numbers. The other numbers are odd.

Here's the introductory exercise:

a. (Display:)   [124:5A]

1 ② 3 ④ 5 ⑥ 7 ⑧ 9 ⑩

11 ⑫ 13 ⑭ 15 ⑯ 17 ⑱ 19 ⑳

The circled numbers are the numbers for counting by twos to 20.
b. Here's a rule about numbers: The numbers for counting by twos are called **even** numbers. What are they called? (Signal.) *Even numbers.*
• The other numbers are called **odd** numbers. What are they called? (Signal.) *Odd numbers.*
• Say the even numbers to 20. Get ready. (Signal.) *2, 4, 6, 8, 10, 12, 14, 16, 18, 20.*
c. My turn to say the odd numbers to 20. (**Point and count.**) 1, 3, 5, 7, 9, 11, 13, 15, 17, 19.
• Your turn: Say the odd numbers to 20. Get ready. (Signal.) *1, 3, 5, 7, 9, 11, 13, 15, 17, 19.*
d. Remember, even numbers are the numbers for counting by twos.
e. I'll say numbers. Tell me if they are even or odd.
• 16. Even or odd? (Signal.) *Even.* Yes, 16 is a number for counting by twos, so it's even.
• 13. Even or odd? (Signal.) *Odd.*
• 20. Even or odd? (Signal.) *Even.*
• 7. Even or odd? (Signal.) *Odd.*
• 11. Even or odd? (Signal.) *Odd.*
• 8. Even or odd? (Signal.) *Even.*
(Repeat until firm.)

Lesson 124, Exercise 5

On Lesson 126, students identify numbers as odd or even. Here's the exercise:

a. You've learned the names odd and even.
- What do we call numbers for counting by twos? (Signal.) *Even.*
- What do we call other numbers? (Signal.) *Odd.*

b. I'll say numbers. Tell me if they are odd or even.
- 26. Odd or even? (Signal.) *Even.*
- 30. Odd or even? (Signal.) *Even.*
- 50. Odd or even? (Signal.) *Even.*
- 53. Odd or even? (Signal.) *Odd.*
- 11. Odd or even? (Signal.) *Odd.*
- 14. Odd or even? (Signal.) *Even.*
- 17. Odd or even? (Signal.) *Odd.*
(Repeat until firm.)

━━━━━ **WORKBOOK PRACTICE** ━━━━━

a. Find part 2 in your workbook. ✔
(Teacher reference:)

a. 21 odd / even		e. 100 odd / even	
b. 19 odd / even		f. 60 odd / even	
c. 42 odd / even		g. 5 odd / even	
d. 79 odd / even			

b. Touch item A. ✔
It shows a number and two names.
- The first name is odd. What's the other name? (Signal.) *Even.*
- Number A is 21. Circle the word to show if 21 is odd or even. ✔
- Everybody, what word did you circle? (Signal.) *Odd.*

c. Circle words for the rest of the items in part 2. (Observe students and give feedback.)

d. Check your work.
I'll read each number. Tell me the word you circled.
- Item B. 19. (Signal.) *Odd.*
- Item C. 42. (Signal.) *Even.*
- Item D. 79. (Signal.) *Odd.*
- Item E. 100. (Signal.) *Even.*
- Item F. 60. (Signal.) *Even.*
- Item G. 5. (Signal.) *Odd.*

Lesson 126, Exercise 3

Teaching Note: Students should not have trouble discriminating odd and even. If they falter on numbers like 79, say the number with the nine stressed: seventy NINE. Then say:

"Is 9 even?"

"So is 79 even?"

The last odd-even activity directs students to create groups of two by making slash lines after every second object. If the last slash line comes after the last object, the entire group has an even number. If the slash occurs before the last object, the group has an odd number.

Here are the items from Lesson 128:

Workbook Lesson 128, Part 3

NUMBER OPERATIONS AND BASE TEN (NBT)

Common Core State Standards

2.NBT 1a: 100 can be thought of as a bundle of ten tens—called a "hundred."

2.NBT 7: Add and subtract within 1000, using concrete models or drawings . . . based on place value . . .

CMC Level C meets these standards partly through the place-value track. In addition, there are exercises that appear on Lessons 126–129.

Here's the exercise from Lesson 126:

---

a. (Hand out lined paper.)
- Pencils down. ✔
b. Listen: 3 tens is 30.
- What's 3 tens? (Signal.) *30.*
- What's 6 tens? (Signal.) *60.*
- What's 8 tens? (Signal.) *80.*
c. What's 9 tens? (Signal.) *90.*
- So what's 10 tens? (Signal.) *100.*
- Yes, 10 tens is 100.
  Say that statement. (Signal.) *10 tens is 100.*
d. What's 11 tens? (Signal.) *110.*
- What's 12 tens? (Signal.) *120.*
- What's 15 tens? (Signal.) *150.*
- What's 18 tens? (Signal.) *180.*
  (Repeat until firm.)
e. You're going to write numbers on your lined paper. I'll say the number of tens. You'll write the number it equals.
- What will you write for 10 tens? (Signal.) *100.*
- What will you write for 13 tens? (Signal.) *130.*
f. Number A is 9 tens. Write the number for 9 tens. ✔
  (Display:)   [126:7A]

  | a. 90 |

  Here's what you should have. 9 tens is 90.
- Number B is 10 tens. Write the number for 10 tens. ✔
  (Display:)   [126:7B]

  | b. 100 |

  Here's what you should have. 10 tens is 100.
- Number C is 14 tens. Write the number for 14 tens. ✔
  (Display:)   [126:7C]

  | c. 140 |

  Here's what you should have. 14 tens is 140.
- Number D is 17 tens. Write the number for 17 tens. ✔
  (Display:)   [126:7D]

  | d. 170 |

  Here's what you should have. 17 tens is 170.

Lesson 126, Exercise 7

On later lessons, students work with groups of ten and bundles of 100. Here's part of the exercise from Lesson 128:

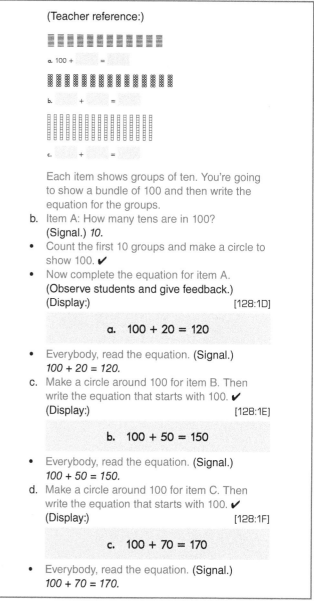

Each item shows groups of ten. You're going to show a bundle of 100 and then write the equation for the groups.
b. Item A: How many tens are in 100? (Signal.) *10.*
- Count the first 10 groups and make a circle to show 100. ✔
- Now complete the equation for item A. (Observe students and give feedback.)
  (Display:)   [128:1D]

  | a. 100 + 20 = 120 |

- Everybody, read the equation. (Signal.) *100 + 20 = 120.*
c. Make a circle around 100 for item B. Then write the equation that starts with 100. ✔
  (Display:)   [128:1E]

  | b. 100 + 50 = 150 |

- Everybody, read the equation. (Signal.) *100 + 50 = 150.*
d. Make a circle around 100 for item C. Then write the equation that starts with 100. ✔
  (Display:)   [128:1F]

  | c. 100 + 70 = 170 |

- Everybody, read the equation. (Signal.) *100 + 70 = 170.*

from Lesson 128, Exercise 1

Another exercise in the same lesson focuses on addition and subtraction. Students are shown how to translate bundles of 100 and groups of ten to work addition and subtraction problems. For subtraction they cross out bundles of hundreds and groups of ten.

Here is part of the exercise:

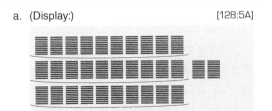

a. (Display:) [128:5A]

This picture shows groups of 10.

b. You're going to work addition problems and subtraction problems that show bundles of 100 and groups of 10.
- How many bundles of 100 are in this picture? (Signal.) *3.*
- How many groups of 10 are left over? (Signal.) *2.*
- What's 300 plus 20? (Signal.) *320.*
- So what number is shown in this picture? (Signal.) *320.*
Yes, 320.
(Add to show:) [128:5B]

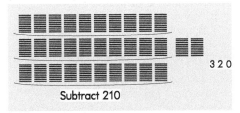

320

Subtract 210

c. I'll read the directions: **Subtract 210.**
- To subtract 210, how many hundreds do I cross out? (Signal.) *2.*
Yes, 2.
(Add to show:) [128:5C]

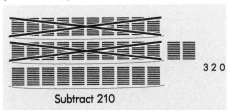

320

Subtract 210

We subtracted 200. Now we have to subtract 1 ten.
- How many tens do I cross out? (Signal.) *1.*
(Add to show:) [128:5D]

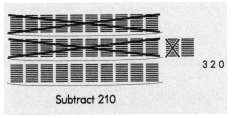

320

Subtract 210

d. The bundles and groups that are not crossed out show how many we end up with.
- How many hundreds are not crossed out? (Signal.) *1.*
- How many tens are not crossed out? (Signal.) *1.*
We end up with 100 and 1 ten. That's 110.
Here's the complete problem.
(Add to show:) [128:5E]

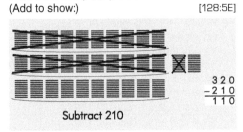

$$\begin{array}{r} 320 \\ -210 \\ \hline 110 \end{array}$$

Subtract 210

e. Let's see if that's right.
- Say the problem and answer for the ones. (Signal.) *0 – 0 = 0.*
- Say the problem and answer for the tens. (Signal.) *2 – 1 = 1.*
- Say the problem and the answer for the hundreds. (Signal.) *3 – 2 = 1.*
The equation agrees with the picture. There are 110 lines that are not crossed out.

from Lesson 128, Exercise 5

2.NBT 3:   Read and write numbers to 1000
using . . . number names . . .

For this standard, students write numbers for
number names. Here's the exercise from Lesson 71:

a. (Display:)   [71:6A]

    thirty-nine

    four hundred six

    eighty-five

    seventeen

    seventy

    forty

    one hundred twenty

    These are number names.
b. My turn to read each number. (Point and read:)
    39, 406, 85, 17, 70, 40, 120.
    • Your turn to read these names. (Point to each
      name as students read:) *39, 406, 85, 17, 70,
      40, 120.*
    (Repeat until firm.)

━━━━━━━ WORKBOOK PRACTICE ━━━━━━━

a. Find part 3 in your workbook. ✔
    (Teacher reference:)

    a. one hundred twenty _____   e. one hundred thirty _____

    b. thirty-eight _____   f. two hundred six _____

    c. sixteen _____   g. thirty-four _____

    d. sixty _____

    You're going to read these number names.
    Then you'll write the number for each name.
b. Touch and read name A. (Signal.) *120.*
    • Read name B. (Signal.) *38.*
    • Name C. (Signal.) *16.*
    • Name D. (Signal.) *60.*
    • Name E. (Signal.) *130.*
    • Name F. (Signal.) *206.*
    • Name G. (Signal.) *34.*
    (Repeat until firm.)
c. Write numbers for the names.
    (Observe students and give feedback.)
d. Check your work.
    (Display:)   [71:6B]

    | a. 120 | e. 130 |
    | b. 38 | f. 206 |
    | c. 16 | g. 34 |
    | d. 60 | |

    Here's what you should have.

Lesson 71, Exercise 6

2.NBT 8:   Mentally add . . . 100 to a given
number 100–900, and mentally
subtract . . . 100 from a given number
100–900.

Students regularly do mental math that meets
this standard. Students add and subtract hundreds
numbers such as 400 + 200 and 700 – 100. They
also add or subtract related hundreds numbers.

Here are parts of the exercises from Lessons 115
and 116:

f. Listen: What's 300 plus 100? (Signal.) *400.*
    • What's 200 plus 300? (Signal.) *500.*
    • What's 320 plus 100? (Signal.) *420.*
    • What's 852 plus 100? (Signal.) *952.*
    • What's 536 plus 100? (Signal.) *636.*
    (Repeat until firm.)

From Lesson 115

e. Listen: What's 400 plus 100? (Signal.) *500.*
    • What's 600 plus 300? (Signal.) *900.*
    • What's 500 plus 200? (Signal.) *700.*
    • What's 500 minus 100? (Signal.) *400.*
f. What's 520 minus 100? (Signal.) *420.*
    • What's 318 minus 100? (Signal.) *218.*
    • What's 790 minus 100? (Signal.) *690.*

From Lesson 116

Measurement and Data (MD)

Common Core State Standards

2.MD 2: Measure the length of an object twice, using length units of different lengths for the two measurements . . .

2.MD 3: Estimate lengths using units of . . . feet . . . and meters.

Students perform measurements that satisfy this standard starting on Lessons 85–87. Here's the exercise from Lesson 86:

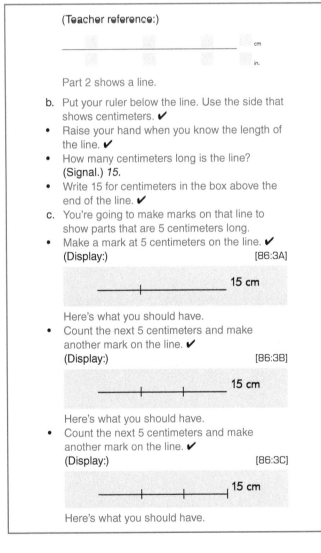

(Teacher reference:)

Part 2 shows a line.

b. Put your ruler below the line. Use the side that shows centimeters. ✔
- Raise your hand when you know the length of the line. ✔
- How many centimeters long is the line? (Signal.) *15.*
- Write 15 for centimeters in the box above the end of the line. ✔

c. You're going to make marks on that line to show parts that are 5 centimeters long.
- Make a mark at 5 centimeters on the line. ✔ (Display:) [86:3A]

Here's what you should have.
- Count the next 5 centimeters and make another mark on the line. ✔ (Display:) [86:3B]

Here's what you should have.
- Count the next 5 centimeters and make another mark on the line. ✔ (Display:) [86:3C]

Here's what you should have.

d. How long is the first part? (Signal.) *5 centimeters.*
- How long is the second part? (Signal.) *5 centimeters.*
- How long is the third part? (Signal.) *5 centimeters.*
Each part is 5 centimeters long.
- Write 5 above each part. ✔ (Display:) [86:3D]

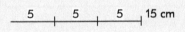

Here's what you should have.
Each part is 5 centimeters long.
- So about how many inches is each part? (Signal.) *2.*
- Write 2 below each part. ✔ (Display:) [86:3E]

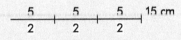

Here's what you should have.

e. Raise your hand when you know about how many inches the whole line is. Add 2 plus 2 plus 2. ✔
- What's 2 plus 2 plus 2? (Signal.) *6.*
- So about how many inches is the whole line? (Signal.) *6.*
- Write 6 below the end of the line. ✔

f. Measure the line in inches. Raise your hand when you know the number of inches that is closest to the end of the line. ✔
- Everybody, what's the number of inches that is closest to the end of the line? (Signal.) *6.*
So 15 centimeters is about 6 inches.
- Say the statement about 15 centimeters. (Signal.) *15 centimeters is about 6 inches.*

from Lesson 86, Exercise 3

This exercise follows lessons in which students express the relationship between inches and centimeters—5 centimeters is about 2 inches.

On other lessons, students measure lines in inches and estimate to the nearest centimeter.

On Lessons 125–127, students work similar problems involving meters and feet. They learn the relationship that one meter is about three feet. The first tasks are of the form "About how many feet is 4 meters?"

On Lesson 127, they work with pictures that represent the lengths of different objects. Here's part of the exercise.

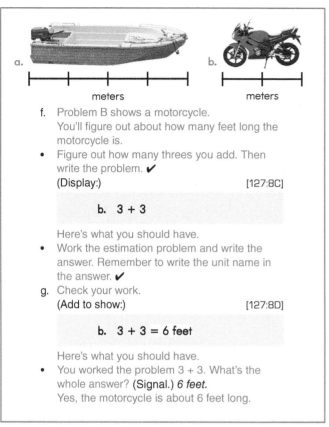

f. Problem B shows a motorcycle. You'll figure out about how many feet long the motorcycle is.
- Figure out how many threes you add. Then write the problem. ✔
(Display:) [127:8C]

> **b.** 3 + 3

Here's what you should have.
- Work the estimation problem and write the answer. Remember to write the unit name in the answer. ✔
g. Check your work.
(Add to show:) [127:8D]

> **b.** 3 + 3 = 6 feet

Here's what you should have.
- You worked the problem 3 + 3. What's the whole answer? (Signal.) *6 feet.*
Yes, the motorcycle is about 6 feet long.

from Lesson 127, Exercise 8

Common Core State Standards

2.MD 7: Tell and write time from . . . digital clocks . . .

The time-telling track starts on Lesson 79. When students write the time shown on analog clocks, they write it in digital form, for instance 9:25.

On Lessons 98 and 99, they match the time shown on digital clocks to that shown on analog clocks.

Here's the exercise from Lesson 98:

a. I'll show you times. You'll tell me the letter of the clock that has the same time.
(Display:) [98:8A]

b. (Add to show:) [98:8B]

- (Point.) Read this time. (Signal.) *7 forty.*
- Raise your hand when you know which clock has the time of 7:40. ✔
- Everybody, which clock has the same time? (Signal.) *C.*
c. (Change to show:) [98:8C]

- What time does this clock show? (Signal.) *7 fifteen.*
- Raise your hand when you know which clock has the time of 7:15. ✔
- Everybody, which clock has the same time? (Signal.) *B.*
d. (Change to show:) [98:8D]

- What time does this clock show? (Signal.) *7 o'clock.*
- Raise your hand when you know which clock has the time of 7 o'clock. ✔
- Everybody, which clock has the same time? (Signal.) *D.*
e. (Change to show:) [98:8E]

- What time does this clock show? (Signal.) *7 thirty-two.*
- Raise your hand when you know which clock has the time of 7:32. ✔
- Everybody, which clock has the same time? (Signal.) *A.*

Lesson 98, Exercise 8

Geometry (G)

Common Core State Standards

2.G 1: Recognize and draw shapes having specified attributes, such as a given number of angles . . .

In the Geometry track students identify shapes and describe their attributes (number of angles and number of sides). On Lessons 128 and 129 students also draw shapes based on specific attributes.

Here's the exercise from Lesson 129:

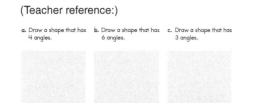

(Teacher reference:)

a. Draw a shape that has 4 angles. b. Draw a shape that has 6 angles. c. Draw a shape that has 3 angles.

You're going to draw a shape for each item.

b. Touch A. ✔
The directions say: **Draw a shape that has 4 angles.** The angles don't have to be the same size.

• Draw your shape.
(Observe students and give feedback.)

c. Check your work.

• Hold up your workbook and show me the shape you drew. ✔

• Did everybody make a quadrilateral? [Students respond.]

d. Touch B. ✔
The directions say: **Draw a shape that has 6 angles.** This is tough. Don't make your sides too long, or you won't be able to make your shape in the space on your worksheet.

• Draw a shape with straight sides and 6 angles.
(Observe students and give feedback.)

e. Check your work.

• Hold up your workbook and show me the shape you drew. ✔

• What's the name of the shape you drew? (Signal.) *Hexagon.*

f. Touch C. ✔
The directions say: **Draw a shape that has 3 angles.**

• Draw your shape.
(Observe students and give feedback.)

g. Check your work.

• Hold up your workbook and show me the shape you drew. ✔

• What's the name of the shape you drew? (Signal.) *Triangle.*

Lesson 129, Exercise 3

Common Core State Standards

2.G 2: Partition a rectangle into rows and columns of same-size squares and count to find the total number of them.

Work on this standard begins on Lesson 47.

Prior to this lesson, students have found the area of rectangles. The procedure they follow is to count the squares in each row and count the number of rows.

They then write a times equation and solve it.

Here's the exercise that introduces the task of partitioning a rectangle into equal sized squares:

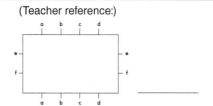

(Teacher reference:)

You're going to make squares inside this rectangle.

b. Touch **A** on the top of the rectangle. ✔

• Touch **A** on the bottom of the rectangle. ✔

• Very carefully put your ruler so the long edge is touching both marks for A.
(Observe students and give feedback.)

• Draw a line from one mark to the other mark.
(Observe students and give feedback.)

c. Now draw a line for the **B**s. Put your ruler so the edge is touching both marks for B. Then make the line.
(Observe students and give feedback.)

• Now draw the lines for the **C**s and the **D**s.
(Observe students and give feedback.)

d. Now you're going to draw the lines for the **E**s and the **F**s. Put your ruler so the edge is touching both marks for **E.** ✔

• Draw the line for the **E**s. Then draw lines for the **F**s.
(Observe students and give feedback.)
You should have squares in your rectangle.

e. On the line next to the rectangle, write the problem and the answer for figuring out the number of squares in your rectangle. Don't forget to make the dots. ✔
(Display:)   [47:4A]

$$5 \times 3 = 15$$
$$ \cdot\cdot\cdot$$

Here's what you should have. There are 5 squares in each row. There are 3 rows.

• Raise your hand if you got it right. ✔

f. Now count the squares and see if there are 15 squares. ✔

• How many squares are there? (Signal.) *15.*

Lesson 47, Exercise 4

2.G 3:   . . . Recognize that equal shares of identical wholes need not have the same shape.

Here's the exercise from Lesson 124 in the regular program:

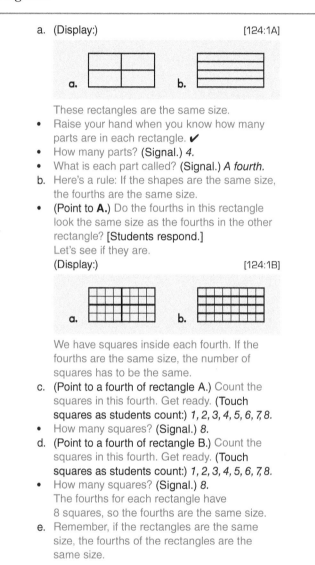

a. (Display:)   [124:1A]

   a. b.

   These rectangles are the same size.
   • Raise your hand when you know how many
     parts are in each rectangle. ✔
   • How many parts? (Signal.) *4.*
   • What is each part called? (Signal.) *A fourth.*
   b. Here's a rule: If the shapes are the same size,
     the fourths are the same size.
   • (Point to **A.**) Do the fourths in this rectangle
     look the same size as the fourths in the other
     rectangle? [Students respond.]
     Let's see if they are.
     (Display:)   [124:1B]

   a. b.

   We have squares inside each fourth. If the
   fourths are the same size, the number of
   squares has to be the same.
   c. (Point to a fourth of rectangle A.) Count the
     squares in this fourth. Get ready. (Touch
     squares as students count:) *1, 2, 3, 4, 5, 6, 7, 8.*
   • How many squares? (Signal.) *8.*
   d. (Point to a fourth of rectangle B.) Count the
     squares in this fourth. Get ready. (Touch
     squares as students count:) *1, 2, 3, 4, 5, 6, 7, 8.*
   • How many squares? (Signal.) *8.*
     The fourths for each rectangle have
     8 squares, so the fourths are the same size.
   e. Remember, if the rectangles are the same
     size, the fourths of the rectangles are the
     same size.

Lesson 124, Exercise 1

Appendix A

Placement Test

Appendix A: Placement Test

The Placement Test provides for three outcomes:

- The student lacks the necessary skills to place in *CMC Level C*.
- The student places at Lesson 1 of *CMC Level C*.
- The student places at Lesson 11 of *CMC Level C*.

The test has two sections: 1 and 2.

Students who have not gone through *CMC Level B* should take Section 1. Results of this section will determine whether each student has the necessary skills to place at Lesson 1. Students who have gone through *CMC Level B* should take Section 2. Results of this section will determine whether a student places at Lesson 1 or Lesson 11 of *Level C*.

If possible, present the Placement Test on the first day of instruction. Pass out a test to each student. Present the wording in the test Administration Directions script.

Note: What you say is shown in blue type. When observing students, make sure that they are working on the correct part of the test. Do not prompt them in a way that would let them know the answer to an item.

Reproducible copies of the test appear on pages 139–141 of this guide.

CONNECTING MATH CONCEPTS— LEVEL C

PLACEMENT TEST, Section 1

Administration Directions

a. (Hand out Placement Test, Section 1.)
- Write your name on the top line. (Check student responses.)

b. Touch Part 1. ✔

Part 1

a. 6 – 1 = ____	d. 10 + 1 = ____
b. 5 – 1 = ____	e. 7 – 2 = ____
c. 2 + 6 = ____	f. 9 + 0 = ____

You'll read each problem.
- Touch A. ✔
  Read the problem. (Signal.) *6 – 1.*
- Touch B. ✔
  Read the problem. (Signal.) *5 – 1.*
- Touch C. ✔
  Read the problem. (Signal.) *2 + 6.*
- Touch D. ✔
  Read the problem. (Signal.) *10 + 1.*
- Touch E. ✔
  Read the problem. (Signal.) *7 – 2.*
- Touch F. ✔
  Read the problem. (Signal.) *9 + 0.*

c. Write answers to all the problems in Part 1. If you don't know an answer, do not look at anybody else's paper. Pencils down when you're finished.
   (Observe students but do not give feedback.)

d. Touch Part 2. ✔

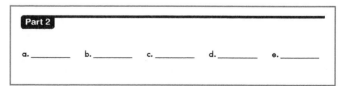

Part 2

a. _____ b. _____ c. _____ d. _____ e. _____

I'll say numbers. You'll write them.
- Touch A. ✔
  52. What number? (Signal.) *52.*
  Write it. ✔

- Touch B. ✔
  71. What number? (Signal.) *71.*
  Write it. ✔
- Touch C. ✔
  12. What number? (Signal.) *12.*
  Write it. ✔
- Touch D. ✔
  17. What number? (Signal.) *17.*
  Write it. ✔
- Touch E. ✔
  80. What number? (Signal.) *80.*
  Write it. ✔
e. Touch Part 3. ✔

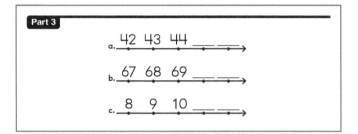

You'll write a number in each blank.
The numbers you'll write are the numbers
you say when you count.
- Touch A. ✔
  The numbers that are shown are 42,
  43, 44.
  Write the next two numbers that come
  after 44.
  (Observe students.)
- Touch B. ✔
  The numbers that are shown are 67,
  68, 69.
  Write the next two numbers.
  (Observe students.)
- Touch C. ✔
  The numbers that are shown are 8, 9, 10.
  Write the next two numbers.
  (Observe students.)
f. (Collect and score Placement Test,
  Section 1.)

Students who make zero to 3 errors should
take Section 2 of the Placement Test.
Students who make more than 3 errors
lack the entry-level skills for this level of the
program.

CONNECTING MATH CONCEPTS— LEVEL C

PLACEMENT TEST, SECTION 2

Administration Directions

a. (Hand out Placement Test, Section 2.)
- Write your name on the top line.
  (Check student responses.)
b. Touch Part 1. ✔

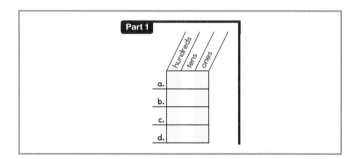

You're going to write numbers. Some are
ones numbers, some are tens numbers,
and some are hundreds numbers.
- Touch A. ✔
  5 hundred 13. What number? (Signal.) *513.*
  Write it. ✔
- Touch B. ✔
  7. What number? (Signal.) *7.*
  Write it. ✔
- Touch C. ✔
  30. What number? (Signal.) *30.*
  Write it. ✔
- Touch D. ✔
  6 hundred 5. What number? (Signal.) *605.*
  Write it. ✔

c. Touch Part 2. ✔

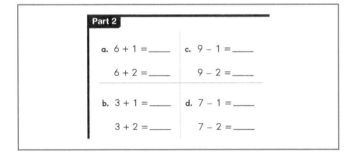

Some problems plus 1 and plus 2. Some problems minus 1 and minus 2.

- Read each problem to yourself and write the answer. ✔

d. Touch Part 3. ✔

- Fill in the missing numbers. ✔

e. Touch part 4. ✔

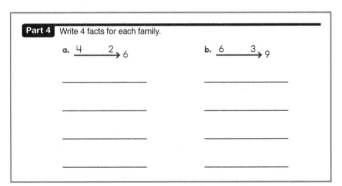

- Write four facts for each family. ✔

f. Touch Part 5. ✔

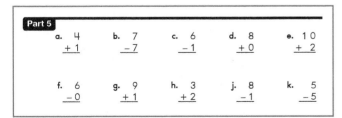

- Read each problem to yourself and write the answer. ✔

g. Touch Part 6. ✔

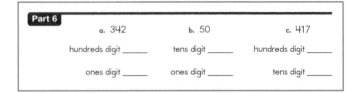

- Touch the number for A. ✔
  You'll write the hundreds digit and the ones digit.
  Listen again: Hundreds digit and ones digit. Write them. ✔
- Touch the number for B. ✔
  You'll write the tens digit and the ones digit.
  Listen again: Tens digit and ones digit. Write them. ✔
- Touch the number for C. ✔
  You'll write the hundreds digit and the tens digit.
  Listen again: Hundreds digit and tens digit. Write them. ✔

h. Touch Part 7. ✔

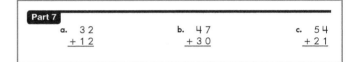

- Work each problem. ✔
  (Collect and score Placement Test, Section 2.)

SCORING THE TEST

(Students who make zero to 10 errors begin instruction at Lesson 11. Students who make more than 10 errors begin instruction at Lesson 1.)

errors

CMC Level C Placement Test Section 1 Name ____________

Part 1

a. 6 – 1 = ____

b. 5 – 1 = ____

c. 2 + 6 = ____

d. 10 + 1 = ____

e. 7 – 2 = ____

f. 9 + 0 = ____

Part 2

a. ________ b. ________ c. ________ d. ________ e. ________

Part 3

a. 42 43 44 ___ ___ →

b. 67 68 69 ___ ___ →

c. 8 9 10 ___ ___ →

CMC Level C Placement Test Section 2   Name _____

Part 1

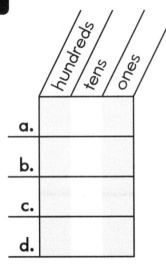

hundreds tens ones

a.

b.

c.

d.

Part 2

a. $6 + 1 =$ _____

$6 + 2 =$ _____

b. $3 + 1 =$ _____

$3 + 2 =$ _____

c. $9 - 1 =$ _____

$9 - 2 =$ _____

d. $7 - 1 =$ _____

$7 - 2 =$ _____

Part 3   Write the missing numbers.

a. 10 9 ___ ___ ___ ___ 4 ___ ___ ___ ___

b. 20 ___ 18 ___ ___ ___ ___ 13 ___ ___ ___

Part 4   Write 4 facts for each family.

a. 4 ——2——> 6

_____

_____

_____

_____

b. 6 ——3——> 9

_____

_____

_____

_____

Part 5

a. 4 + 1	b. 7 − 7	c. 6 − 1	d. 8 + 0	e. 1 0 + 2

f. 6 − 0	g. 9 + 1	h. 3 + 2	j. 8 − 1	k. 5 − 5

Part 6

a. 342

hundreds digit _____

ones digit _____

b. 50

tens digit _____

ones digit _____

c. 417

hundreds digit _____

tens digit _____

Part 7

a. 3 2 + 1 2	b. 4 7 + 3 0	c. 5 4 + 2 1

CMC Level C Placement Test Answer Key, Section 1

[varies] errors

CMC Level C Placement Test Section 1   Name _____

Part 1

a. $6 - 1 = \underline{5}$ d. $10 + 1 = \underline{11}$

b. $5 - 1 = \underline{4}$ e. $7 - 2 = \underline{5}$

c. $2 + 6 = \underline{8}$ f. $9 + 0 = \underline{9}$

Part 2

a. _52_   b. _71_   c. _12_   d. _17_   e. _80_

Part 3

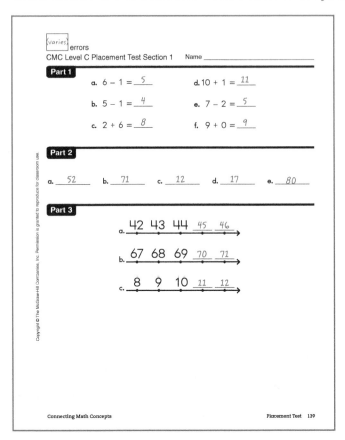

a. 42 43 44 _45_ _46_

b. 67 68 69 _70_ _71_

c. 8 9 10 _11_ _12_

Connecting Math Concepts   Placement Test 139

CMC Level C Placement Test Answer Key, Section 2

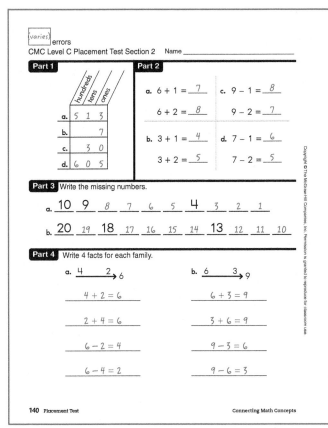

[varies] errors

CMC Level C Placement Test Section 2   Name _____

Part 1

	hundreds	tens	ones
a.	5	1	3
b.			7
c.		3	0
d.	6	0	5

Part 2

a. $6 + 1 = \underline{7}$ c. $9 - 1 = \underline{8}$

$6 + 2 = \underline{8}$ $9 - 2 = \underline{7}$

b. $3 + 1 = \underline{4}$ d. $7 - 1 = \underline{6}$

$3 + 2 = \underline{5}$ $7 - 2 = \underline{5}$

Part 3 Write the missing numbers.

a. 10 9 _8_ _7_ _6_ _5_ 4 _3_ _2_ _1_

b. 20 _19_ 18 _17_ _16_ _15_ _14_ 13 _12_ _11_ _10_

Part 4 Write 4 facts for each family.

a. $4 \xrightarrow{\quad 2 \quad} 6$

_____ $4 + 2 = 6$ _____

_____ $2 + 4 = 6$ _____

_____ $6 - 2 = 4$ _____

_____ $6 - 4 = 2$ _____

b. $6 \xrightarrow{\quad 3 \quad} 9$

_____ $6 + 3 = 9$ _____

_____ $3 + 6 = 9$ _____

_____ $9 - 3 = 6$ _____

_____ $9 - 6 = 3$ _____

140 Placement Test   Connecting Math Concepts

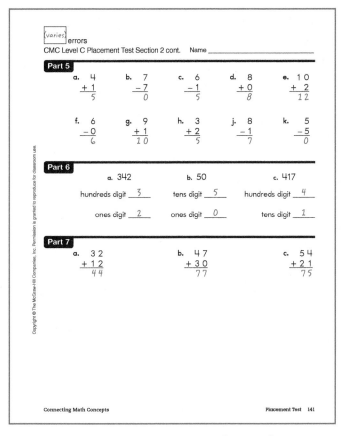

[varies] errors

CMC Level C Placement Test Section 2 cont.   Name _____

Part 5

a. $\begin{array}{r} 4 \\ +1 \\ \hline 5 \end{array}$ b. $\begin{array}{r} 7 \\ -7 \\ \hline 0 \end{array}$ c. $\begin{array}{r} 6 \\ -1 \\ \hline 5 \end{array}$ d. $\begin{array}{r} 8 \\ +0 \\ \hline 8 \end{array}$ e. $\begin{array}{r} 10 \\ +2 \\ \hline 12 \end{array}$

f. $\begin{array}{r} 6 \\ -0 \\ \hline 6 \end{array}$ g. $\begin{array}{r} 9 \\ +1 \\ \hline 10 \end{array}$ h. $\begin{array}{r} 3 \\ +2 \\ \hline 5 \end{array}$ j. $\begin{array}{r} 8 \\ -1 \\ \hline 7 \end{array}$ k. $\begin{array}{r} 5 \\ -5 \\ \hline 0 \end{array}$

Part 6

a. 342   b. 50   c. 417

hundreds digit _3_   tens digit _5_   hundreds digit _4_

ones digit _2_   ones digit _0_   tens digit _1_

Part 7

a. $\begin{array}{r} 3\,2 \\ +1\,2 \\ \hline 4\,4 \end{array}$ b. $\begin{array}{r} 4\,7 \\ +3\,0 \\ \hline 7\,7 \end{array}$ c. $\begin{array}{r} 5\,4 \\ +2\,1 \\ \hline 7\,5 \end{array}$

Connecting Math Concepts   Placement Test 141

142 *Teacher's Guide*   **Connecting Math Concepts**

Appendix B

Reproducible Mastery Test
Summary Sheets

Remedy Summary—Group Summary of Test Performance

Note: Test remedies are included in the *Student Assessment Book.* Percent Tables are provided in the *Answer Key.*

Name	Test 1 Check parts not passed							Total %	Test 2 Check parts not passed												Total %	Test 3 Check parts not passed									Total %
	1	2	3	4	5	6	7		1	2	3	4	5	6	7	8	9	10	11	12		1	2	3	4	5	6	7	8	9	
1.																															
2.																															
3.																															
4.																															
5.																															
6.																															
7.																															
8.																															
9.																															
10.																															
11.																															
12.																															
13.																															
14.																															
15.																															
16.																															
17.																															
18.																															
19.																															
20.																															
21.																															
22.																															
23.																															
24.																															
25.																															
26.																															
27.																															
28.																															
29.																															
30.																															
Number of students Not Passed = NP																															
Total number of students = T																															
Remedy needed if NP/T = 25% or more																															

Connecting Math Concepts

Remedy Summary—Group Summary of Test Performance

Note: Test remedies are included in the *Student Assessment Book*. Percent Tables are provided in the *Answer Key*.

Test 4

Name	Check parts not passed											Total %
	1	2	3	4	5	6	7	8	9	10	11	
1.												
2.												
3.												
4.												
5.												
6.												
7.												
8.												
9.												
10.												
11.												
12.												
13.												
14.												
15.												
16.												
17.												
18.												
19.												
20.												
21.												
22.												
23.												
24.												
25.												
26.												
27.												
28.												
29.												
30.												
Number of students Not Passed = NP												
Total number of students = T												
Remedy needed if NP/T = 25% or more												

Test 5

Check parts not passed									Total %
1	2	3	4	5	6	7	8	9	

Test 6

Check parts not passed							Total %
1	2	3	4	5	6	7	

Remedy Summary—Group Summary of Test Performance

Note: Test remedies are included in the *Student Assessment Book.* Percent Tables are provided in the *Answer Key.*

Name	Test 7 Check parts not passed							Total %	Test 8 Check parts not passed							Total %	Test 9 Check parts not passed									Total %
	1	2	3	4	5	6	7		1	2	3	4	5	6	7		1	2	3	4	5	6	7	8	9	
1.																										
2.																										
3.																										
4.																										
5.																										
6.																										
7.																										
8.																										
9.																										
10.																										
11.																										
12.																										
13.																										
14.																										
15.																										
16.																										
17.																										
18.																										
19.																										
20.																										
21.																										
22.																										
23.																										
24.																										
25.																										
26.																										
27.																										
28.																										
29.																										
30.																										
Number of students Not Passed = NP																										
Total number of students = T																										
Remedy needed if NP/T = 25% or more																										

Remedy Summary—Group Summary of Test Performance

Note: Test remedies are included in the *Student Assessment Book*. Percent Tables are provided in the *Answer Key*.

	Test 10								Test 11											Test 12						
Name	Check parts not passed							Total %	Check parts not passed										Total %	Check parts not passed						Total %
	1	2	3	4	5	6	7		1	2	3	4	5	6	7	8	9	10		1	2	3	4	5	6	
1.																										
2.																										
3.																										
4.																										
5.																										
6.																										
7.																										
8.																										
9.																										
10.																										
11.																										
12.																										
13.																										
14.																										
15.																										
16.																										
17.																										
18.																										
19.																										
20.																										
21.																										
22.																										
23.																										
24.																										
25.																										
26.																										
27.																										
28.																										
29.																										
30.																										
Number of students Not Passed = NP																										
Total number of students = T																										
Remedy needed if NP/T = 25% or more																										

Remedy Summary—Group Summary of Test Performance

Test 13

Note: Test remedies are included in the *Student Assessment Book*. Percent Tables are provided in the *Answer Key*.

Name	Check parts not passed														Total %
	1	2	3	4	5	6	7	8	9	10	11	12	13	14	
1.															
2.															
3.															
4.															
5.															
6.															
7.															
8.															
9.															
10.															
11.															
12.															
13.															
14.															
15.															
16.															
17.															
18.															
19.															
20.															
21.															
22.															
23.															
24.															
25.															
26.															
27.															
28.															
29.															
30.															
Number of students Not Passed = NP															
Total number of students = T															
Remedy needed if NP/T = 25% or more															

Connecting Math Concepts

Appendix C

Sample Lessons

- Lesson 44 Presentation Book, Workbook, and Textbook
- Lesson 102 Presentation Book, Workbook, and Textbook

Lesson 44

Note: Students will need a ruler that shows inches and centimeters for Independent Work, Workbook part 4.

EXERCISE 1: COLUMN SUBTRACTION
RENAMING

a. (Display:)   [44:1A]

$$\begin{array}{r} 6\,2 \\ -2\,9 \end{array}$$

- Read the problem. (Signal.) *62 – 29.*
- Read the problem for the ones. (Signal.) *2 – 9.*
- Can you work that problem? (Signal.) *No.*
  So you have to rewrite a number.
- Read the number you'll rewrite. (Signal.) *62.*
b. Say the new place value for 62. (Signal.)
  50 + 12.
  (Add to show:)   [44:1B]

$$\begin{array}{r} ^5\!\!\!\not{6}^{\,1}2 \\ -2\,9 \end{array}$$

- Say the new problem for the ones. (Signal.)
  12 – 9.
  What's the answer? (Signal.) *3.*
- Say the new problem for the tens. (Signal.)
  5 – 2.
  What's the answer? (Signal.) *3.*
  (Add to show:)   [44:1C]

$$\begin{array}{r} ^5\!\!\!\not{6}^{\,1}2 \\ -2\,9 \\ \hline 3\,3 \end{array}$$

I'll read the problem we started with and the answer: 62 – 29 = 33.
- Everybody, read the problem and the answer. (Signal.) *62 – 29 = 33.*
  (Repeat until firm.)

c. (Display:)   [44:1D]

$$\begin{array}{r} 9\,1 \\ -7\,2 \end{array}$$

- Read the problem. (Signal.) *91 – 72.*
- Read the problem for the ones. (Signal.) *1 – 2.*
- Can you work that problem? (Signal.) *No.*
  So you have to rewrite a number.
- Read the number you'll rewrite. (Signal.) *91.*
d. Say the new place value for 91. (Signal.)
  80 + 11.
  (Add to show:)   [44:1E]

$$\begin{array}{r} ^8\!\!\!\not{9}^{\,1}1 \\ -7\,2 \end{array}$$

- Say the new problem for the ones. (Signal.)
  11 – 2.
  What's the answer? (Signal.) *9.*
- Say the new problem for the tens. (Signal.)
  8 – 7.
  What's the answer? (Signal.) *1.*
  (Add to show:)   [44:1F]

$$\begin{array}{r} ^8\!\!\!\not{9}^{\,1}1 \\ -7\,2 \\ \hline 1\,9 \end{array}$$

- Everybody, read the problem we started with and the answer. (Signal.) *91 – 72 = 19.*
  (Repeat until firm.)

EXERCISE 2: GEOMETRY
2-DIMENSIONAL SHAPES

a. (Display:)   [44:2A]

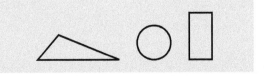

b. These are shapes you know.
- What's the name of the shape that has 4 straight sides? (Signal.) *Rectangle.*
- What's the name of the shape that has 3 straight sides? (Signal.) *Triangle.*
- What's the name of the shape that has no straight sides? (Signal.) *Circle.*

c. Your turn to say the names of these shapes.
- (Point to **triangle.**) What's this shape? (Signal.) *Triangle.*
- (Point to **circle.**) What's this shape? (Signal.) *Circle.*
- (Point to **rectangle.**) What's this shape? (Signal.) *Rectangle.*
  (Repeat until firm.)

━━━━━━━━━ **WORKBOOK PRACTICE** ━━━━━━━━━

a. Open your workbook to Lesson 44 and find part 1. ✔
   (Teacher reference:)

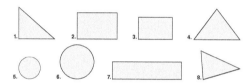

b. You're going to write the letter **R** inside all the rectangles.
- What letter goes inside each rectangle? (Signal.) *R.*
- What letter goes inside each triangle? (Signal.) *T.*
  And **C** goes inside each circle.

c. Write a letter in each shape.
   (Observe students and give feedback.)

d. Check your work.
- Shape 1. What's the name? (Signal.) *Triangle.*
  What letter did you write? (Signal.) *T.*
- Shape 2. What's the name? (Signal.) *Rectangle.*
  What letter did you write? (Signal.) *R.*
- Shape 3. What's the name? (Signal.) *Rectangle.*
  What letter did you write? (Signal.) *R.*

- Shape 4. What's the name? (Signal.) *Triangle.*
  What letter did you write? (Signal.) *T.*
- Shape 5. What's the name? (Signal.) *Circle.*
  What letter did you write? (Signal.) *C.*
- Shape 6. What's the name? (Signal.) *Circle.*
  What letter did you write? (Signal.) *C.*
- Shape 7. What's the name? (Signal.) *Rectangle.*
  What letter did you write? (Signal.) *R.*
- Shape 8. What's the name? (Signal.) *Triangle.*
  What letter did you write? (Signal.) *T.*

EXERCISE 3: FACT REVIEW
SMALL NUMBER OF 10 OR 9

a. (Display:)   [44:3A]

$$\begin{array}{ll} \underset{\longrightarrow}{10 \quad 1} & \underset{\longrightarrow}{9 \quad 1} \\ \underset{\longrightarrow}{10 \quad 2} & \underset{\longrightarrow}{9 \quad 2} \\ \underset{\longrightarrow}{10 \quad 3} & \underset{\longrightarrow}{9 \quad 3} \\ \underset{\longrightarrow}{10 \quad 4} & \underset{\longrightarrow}{9 \quad 4} \\ \underset{\longrightarrow}{10 \quad 5} & \underset{\longrightarrow}{9 \quad 5} \\ \underset{\longrightarrow}{10 \quad 6} & \underset{\longrightarrow}{9 \quad 6} \\ \underset{\longrightarrow}{10 \quad 7} & \underset{\longrightarrow}{9 \quad 7} \\ \underset{\longrightarrow}{10 \quad 8} & \underset{\longrightarrow}{9 \quad 8} \\ \underset{\longrightarrow}{10 \quad 9} & \underset{\longrightarrow}{9 \quad 9} \end{array}$$

b. (Point to **left column.**) These families have a small number of 10.
- (Point to **right column.**) These families have a small number of 9.

c. Here's a rule about families that have a small number of 9: The big number is 1 less than the family that has a small number of 10.

d. (Point to 10 1→.) What's the big number for this family? (Signal.) *11.*
• (Point to 9 1→.) The big number for this family is 1 less than 11. What's the big number? (Signal.) *10.*
e. (Point to 10 2→.) What's the big number for this family? (Signal.) *12.*
• (Point to 9 2→.) What's 1 less than 12? (Signal.) *11.*
• So what's the big number for this family? (Signal.) *11.*
f. (Point to 10 3→.) What's the big number for this family? (Signal.) *13.*
• (Point to 9 3→.) So what's the big number for this family? (Signal.) *12.*
g. (Point to 10 4→.) What's the big number tor this family? (Signal.) *14.*
• (Point to 9 4→.) So what's the big number for this family? (Signal.) *13.*
h. (Point to 10 5→.) What's the big number for this family? (Signal.) *15.*
• (Point to 9 5→.) So what's the big number for this family? (Signal.) *14.*
(Repeat until firm.)
i. This time I'll tell you the small numbers. You'll tell me the big number.
j. (Point to 10 1→.) The small numbers are 10 and 1.
What's the big number? (Signal.) *11.*
• (Point to 9 1→.) The small numbers are 9 and 1.
What's the big number? (Signal.) *10.*
k. (Point to 10 4→.) The small numbers are 10 and 4.
What's the big number? (Signal.) *14.*
• (Point to 9 4→.) The small numbers are 9 and 4.
What's the big number? (Signal.) *13.*
l. (Point to 10 7→.) The small numbers are 10 and 7.
What's the big number? (Signal.) *17.*
• (Point to 9 7→.) The small numbers are 9 and 7.
What's the big number? (Signal.) *16.*
(Repeat until firm.)

WORKBOOK PRACTICE

a. Find part 2 in your workbook. ✔
• Pencils down. ✔
(Teacher reference:)

a. 9 3→ ____   d. 9 7→ ____   g. 9 4→ ____
b. 9 8→ ____   e. 9 9→ ____   h. 9 2→ ____
c. 9 5→ ____   f. 9 1→ ____   i. 9 6→ ____

These are number families that have a small number of 9. You're going to figure out the big number for each family.
b. Family A: What are the small numbers? (Signal.) *9 and 3.*
• Think of the family with 10 and 3. What's the big number for that family? (Signal.) *13.*
• So what's the big number for the family with 9 and 3? (Signal.) *12.*
c. Family B: What are the small numbers? (Signal.) *9 and 8.*
• Think of the family with 10 and 8. What's the big number for that family? (Signal.) *18.*
• So what's the big number for the family with 9 and 8? (Signal.) *17.*
d. Family C: What are the small numbers? (Signal.) *9 and 5.*
• Think of the family with 10 and 5. What's the big number for that family? (Signal.) *15.*
• So what's the big number for the family with 9 and 5? (Signal.) *14.*
e. Family D: What are the small numbers? (Signal.) *9 and 7*
• Think of the family with 10 and 7. What's the big number for that family? (Signal.) *17.*
• So what's the big number for the family with 9 and 7? (Signal.) *16.*
f. Remember, if a family has a small number of 9, think of the family with a small number of 10.
• Write the missing big number for each family. (Observe students and give feedback.)

g. Check your work.
- Family A. What are the small numbers? (Signal.) *9 and 3.*
  What's the big number? (Signal.) *12.*
- Family B. What are the small numbers? (Signal.) *9 and 8.*
  What's the big number? (Signal.) *17.*
- C. What are the small numbers? (Signal.) *9 and 5.*
  What's the big number? (Signal.) *14.*
- D. What are the small numbers? (Signal.) *9 and 7.*
  What's the big number? (Signal.) *16.*
- E. What are the small numbers? (Signal.) *9 and 9.*
  What's the big number? (Signal.) *18.*
- F. What are the small numbers? (Signal.) *9 and 1.*
  What's the big number? (Signal.) *10.*
- G. What are the small numbers? (Signal.) *9 and 4.*
  What's the big number? (Signal.) *13.*
- H. What are the small numbers? (Signal.) *9 and 2.*
  What's the big number? (Signal.) *11.*
- I. What are the small numbers? (Signal.) *9 and 6.*
  What's the big number? (Signal.) *15.*

EXERCISE 4: TRANSITIVITY
WITH LETTERS   REMEDY

a. Find part 3 in your workbook. ✔
   (Teacher reference:)   R Part E

```
a.      H > C          b.      P < J
        C > T                  J < R
      _____             _____

      _____             _____
```

b. Touch problem A. ✔
   These statements tell about H, C, and T.
- Read the top statement. (Signal.) *H is more than C.*
- Read the other statement. (Signal.) *C is more than T.*

c. Write the three letters on the line. Write H as the first letter. Write the signs between the three letters.
   (Observe students and give feedback.)
d. Check your work.
   (Display:)   [44:4A]

   a. H > C > T

   Here's what you should have.
   H is more than C. C is more than T.
- On the line below, write the statement about H and T. ✔
   (Add to show:)   [44:4B]

   a. H > C > T
      H > T

   Here's that you should have. H is more than T.
e. Problem B: These statements tell about P, J, and R.
- Read the first statement. (Signal.) *P is less than J.*
- Read the second statement. (Signal.) *J is less than R.*
f. Write the three letters on the line. Write P as the first letter. Write the signs between the three letters.
   (Observe students and give feedback.)
g. Check your work.
   (Display:)   [44:4C]

   b. P < J < R

   Here's what you should have.
   P is less than J, and J is less than R.
- Below, write the statement about P and R. ✔
   (Add to show:)   [44:4D]

   b. P < J < R
      P < R

   Here's what you should have.
- Read the statement. (Signal.) *P is less than R.*

EXERCISE 5: MULTIPLICATION
ARRAYS

a. (Display:) [44:5A]

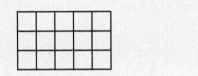

You're going to figure out how many squares there are. But you won't count all of them. You'll work a count-by problem.

b. Count the squares in the top row. Get ready. (Touch as students count:) *1, 2, 3, 4, 5.*
- How many squares are in the top row? (Signal.) *5.*
- So how many squares are in the next row? (Signal.) *5.*
- How many squares are in the bottom row? (Signal.) *5.*

c. Listen: There are 5 squares in each row. So you count by 5.
- What do you count by? (Signal.) *5.*
  (Add to show:) [44:5B]

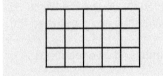

  Yes, you count by 5.

d. Listen: You count by 5 for each row because each row has 5 squares.
  Once more: You count by 5 for each row.
- How many rows are there? (Signal.) *3.*
- So how many times do you count by 5? (Signal.) *3.*
  Yes, count by 5 three times.
- Say that. (Signal.) *Count by 5 three times.*
  (Add to show:) [44:5C]

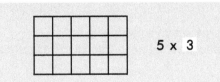

e. I make a dot for each row.
  (Add to show:) [44:5D]

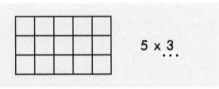

- Everybody, count by 5 three times. (Signal.) *5, 10, 15.*
- What's the answer? (Signal.) *15.*
  (Add to show:) [40:5E]

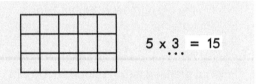

- How many squares are there? (Signal.) *15.*
f. Let's see if the answer is right.
- I'll touch the squares. You count. (Touch squares as students count:) *1, 2, 3, 4, 5, 6, 7, 8, 9, 10, 11, 12, 13, 14, 15.*
- How many squares are there? (Signal.) *15.* You figured out that answer the fast way.

g. (Display:) [44:5F]

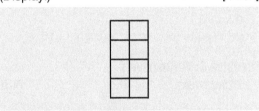

Here's a new problem.
- Raise your hand when you know how many squares are in each row. ✔
- How many squares are in each row? (Signal.) *2.*
- So what do we count by? (Signal.) *2.*
  (Add to show:) [44:5G]

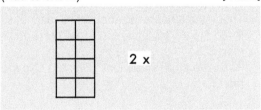

h. Raise your hand when you know how many rows there are. ✔
•   How many rows are there? (Signal.) *4.*
•   So how many times do we count? (Signal.) *4.*
    (Add to show:)   [44:5H]

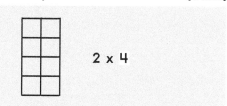

My turn to say the problem we'll work: 2 times 4.
•   Say the problem we'll work. (Signal.) *2 x 4.*
i. I make a dot for each row.
    (Add to show:)   [44:5I]

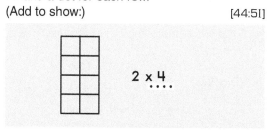

•   Everybody, count by 2 four times. (Signal.) *2, 4, 6, 8.*
•   What's the answer? (Signal.) *8.*
    (Add to show:)   [44:5J]

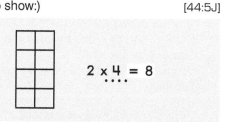

•   Let's count all the squares and see if there are 8 squares. (Touch as students count:) *1, 2, 3, 4, 5, 6, 7, 8.*
    You figured out the answer the fast way.
j. (Display:)   [44:5K]

New problem.
•   Raise your hand when you know how many squares are in each row. ✔

•   How many squares are in each row? (Signal.) *10.*
    (Add to show:)   [44:5L]

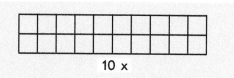

k. Raise your hand when you know how many rows there are. ✔
•   How many rows are there? (Signal.) *2.*
    We'll work the problem 10 times 2.
•   Say the problem we'll work. (Signal.) *10 x 2.*
    (Add to show:)   [44:5M]

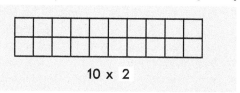

l. I make a dot for each row.
    (Add to show:)   [44:5N]

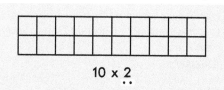

•   Everybody, count by 10 two times. (Signal.) *10, 20.*
•   What's the answer? (Signal.) *20.*
    (Add to show:)   [44:5O]

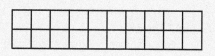

•   Let's count all the squares and see if there are 20 squares. (Touch as students count:) *1, 2, 3, 4, 5, 6, 7, 8, 9, 10, 11, 12, 13, 14, 15, 16, 17, 18, 19, 20.*
    You figured out the answer the fast way.

Connecting Math Concepts

Lesson 44   *309*

EXERCISE 6: COMPARISON VOCABULARY
OLDER/LONGER/TALLER

a. (Display:)   [44:6A]

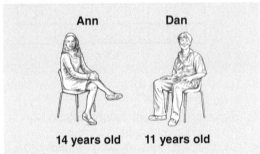

Ann   Dan

14 years old   11 years old

Listen: Ann is 14 years old.
Dan is 11 years old.
- How old is Ann? (Signal.) *14 years old.*
- How old is Dan? (Signal.) *11 years old.*
b. Who is younger—Ann or Dan? (Signal.) *Dan.*
- Who is older—Ann or Dan? (Signal.) *Ann.*
  Yes, Ann is older than Dan.
- Say the sentence. (Signal.) *Ann is older than Dan.*
  Dan is younger than Ann.
- Say the sentence. (Signal.) *Dan is younger than Ann.*

c. (Display:)   [44:6B]

Pete   Mary

6 feet tall   5 feet tall

Listen: Pete is 6 feet tall.
Mary is 5 feet tall.
- How tall is Pete? (Signal.) *6 feet tall.*
- How tall is Mary? (Signal.) *5 feet tall.*
- Who is shorter—Pete or Mary? (Signal.) *Mary.*
- Who is taller? (Signal.) *Pete.*

d. (Display:)   [44:6C]

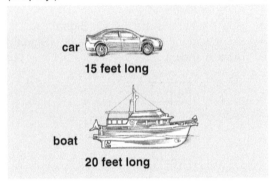

car
15 feet long

boat
20 feet long

Listen: The car is 15 feet long.
The boat is 20 feet long.
- How long is the car? (Signal.) *15 feet.*
- How long is the boat? (Signal.) *20 feet.*
- Which is longer—the car or the boat? (Signal.) *The boat.*
- Which is shorter? (Signal.) *The car.*

e. (Display:)   [44:6D]

> The house is 30 feet tall.
> The garage is 11 feet tall.

(Touch and read:) The house is 30 feet tall.
The garage is 11 feet tall.
- How tall is the house? (Signal.) *30 feet.*
- How tall is the garage? (Signal.) *11 feet.*
- Which is shorter—the house or the garage? (Signal.) *The garage.*
- Which is taller? (Signal.) *The house.*

f. (Display:)   [44:6E]

> The bike was 6 feet long.
> The bus was 60 feet long.

(Touch and read:) The bike was 6 feet long.
The bus was 60 feet long.
- How long was the bike? (Signal.) *6 feet.*
- How long was the bus? (Signal.) *60 feet.*
- Which was longer—the bike or the bus? (Signal.) *The bus.*
- Which was shorter? (Signal.) *The bike.*

EXERCISE 7: ACTION WORD PROBLEMS
START/END NUMBER GIVEN

a. (Hand out textbooks and lined paper.)
- Write your name at the top of your lined paper. ✔
b. Open your textbook to Lesson 44 and find part 1. ✔
(Teacher reference:)

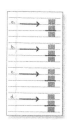

a. A train started out with 60 people.
   Then 31 more people got on the train.
   How many people ended up on the train?

b. A train started out with some people.
   Then 81 more people got on the train.
   The train ended up with 194 people.
   How many people started out on the train?

c. A train started out with some people on it.
   Then 12 people got off the train.
   The train ended up with 100 people.
   How many people started out on the train?

d. A train started out with 90 people.
   Then 40 people got off the train.
   How many people ended up on the train?

You'll make a number family for each problem.
- Write **A** in front of the margin and make a number family arrow. ✔
- Count 4 lines and make the arrow for **B.** Then count lines and make the arrows for **C** and **D.** Your arrows should look just like the picture in part 1.
(Observe students and give feedback.)

c. These are start-end problems. Some of these problems give a number for start. Some give a number for end.
d. Problem A: A train started out with 60 people. Then 31 more people got on the train. How many people ended up on the train?
- Make the number family with letters for start and end and with two numbers. ✔
(Display:)   [44:7A]

$$
\begin{array}{c}
60 \\
\text{a.} \quad \underline{31 \quad \overset{\times}{}} {\to} E
\end{array}
$$

Here's what you should have.
- Touch where you'll write the column problem for family A. ✔
- Write the problem and work it. Remember to put the answer in the number family.
(Observe students and give feedback.)

e. Check your work.
(Add to show:)   [44:7B]

$$
\text{a.} \quad \underline{31 \quad \overset{60}{\overset{\times}{}} \overset{91}{\to} \cancel{E}} \qquad
\begin{array}{r}
31 \\
+60 \\
\hline
91
\end{array}
$$

Here's what you should have.
- Everybody, read the problem and the answer. (Signal.) *31 + 60 = 91.*
- How many people ended up on the train? (Signal.) *91.*
f. Problem B: A train started out with some people. Then 81 more people got on the train. The train ended up with 194 people. How many people started out on the train?
- Make the number family with letters for start and end and with two numbers. ✔
(Display:)   [44:7C]

$$
\text{b.} \quad \underline{81 \quad S \overset{194}{\to} \cancel{E}}
$$

Here's what you should have.
- Touch where you'll write the column problem for family B. ✔
- Write the problem and work it. Remember to put the answer in the number family.
(Observe students and give feedback.)

g. Check your work.
(Add to show:)   [44:7D]

$$
\text{b.} \quad \underline{81 \quad \overset{113}{\overset{\times}{}} \overset{194}{\to} \cancel{E}} \qquad
\begin{array}{r}
194 \\
-81 \\
\hline
113
\end{array}
$$

Here's what you should have.
- Everybody, read the problem and the answer. (Signal.) *194 – 81 = 113.*
- How many people started out on the train? (Signal.) *113.*

h. Problem C: A train started out with some people on it. Then 12 people got off the train. The train ended up with 100 people. How many people started out on the train?
- Make the number family with letters for start and end and with two numbers. ✔
  (Display:)   [44:7E]

100
c. 12   E→S

Here's what you should have.
- Write the problem for family C and work it.
  (Observe students and give feedback.)
i. Check your work.
  (Add to show:)   [44:7F]

100 112
c. 12 E→S

$$\begin{array}{r} 1\,2 \\ +1\,0\,0 \\ \hline 1\,1\,2 \end{array}$$

Here's what you should have.
- Everybody, read the problem and the answer.
  (Signal.) *12 + 100 = 112.*
- How many people started out on the train?
  (Signal.) *112.*
j. Problem D: A train started out with 90 people. Then 40 people got off the train. How many people ended up on the train?
- Make the number family with letters for start and end and with two numbers. ✔
  (Display:)   [44:7G]

90
d. 40 E→S

Here's what you should have.
- Write the problem for family D and work it.
  (Observe students and give feedback.)
k. Check your work.
  (Add to show:)   [44:7H]

50 90
d. 40 E→S

$$\begin{array}{r} 9\,0 \\ -4\,0 \\ \hline 5\,0 \end{array}$$

Here's what you should have.
- Everybody, read the problem and the answer.
  (Signal.) *90 − 40 = 50.*
- How many people ended up on the train?
  (Signal.) *50.*

EXERCISE 8: INDEPENDENT WORK
MULTIPLICATION/MEASUREMENT/INEQUALITY

a. Find part 2 in your textbook. ✔
   (Teacher reference:)

a. 5 x 4 =
b. 4 x 5 =
c. 9 x 6 =

- Count 4 lines and write A. ✔
  The picture shows how to skip lines for these problems. You'll copy and work these problems as part of your independent work.
b. Find part 4 in your workbook. ✔
   (Teacher reference:)

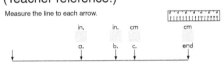

Measure the line to each arrow.

I'll read the directions: **Measure the line to each arrow.**
Be careful. For some, you'll write inches. For some, you'll write centimeters.
- Touch A. Do you measure in inches or centimeters? (Signal.) *Inches.*
- Touch C. Do you measure in inches or centimeters? (Signal.) *Centimeters.*
c. Find part 5 in your workbook. ✔
   (Teacher reference:)

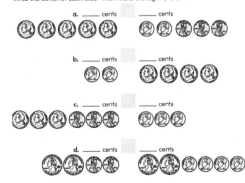

Write the cents for each side. Then make the sign >, <, or =.

I'll read the directions: **Write the cents for each side. Then make the sign more than, less than, or equal to.**
d. You'll work these coin problems as part of your independent work.

Assign Independent Work, Textbook part 2 and Workbook parts 4–11.

Connecting Math Concepts

Lesson 44

Name _____

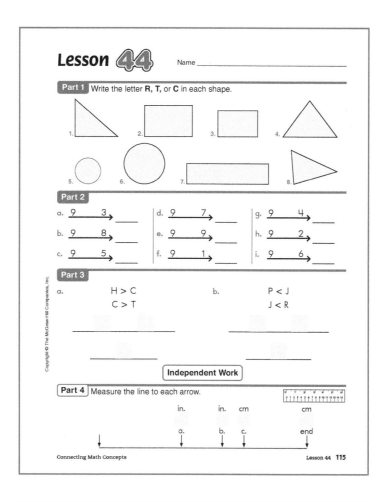

Part 1 Write the letter **R**, **T**, or **C** in each shape.

1. 2. 3. 4.

5. 6. 7. 8.

Part 2

a. 9 ⟶ 3 ____   d. 9 ⟶ 7 ____   g. 9 ⟶ 4 ____

b. 9 ⟶ 8 ____   e. 9 ⟶ 9 ____   h. 9 ⟶ 2 ____

c. 9 ⟶ 5 ____   f. 9 ⟶ 1 ____   i. 9 ⟶ 6 ____

Part 3

a.   H > C   b.   P < J
      C > T   J < R

_____   _____

_____   _____

Independent Work

Part 4 Measure the line to each arrow.

in.   in. cm   cm

a.   b. c.   end

Connecting Math Concepts   Lesson 44 115

Lesson 44

Name _____

Part 5 Write the cents for each side. Then make the sign >, <, or =.

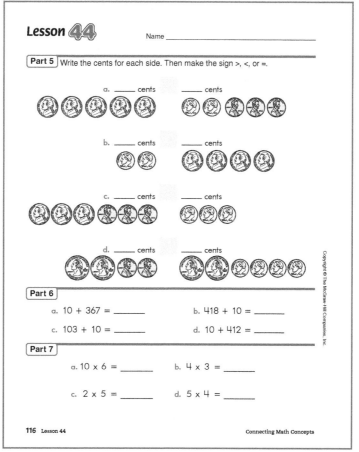

a. ____ cents   ____ cents

b. ____ cents   ____ cents

c. ____ cents   ____ cents

d. ____ cents   ____ cents

Part 6

a. 10 + 367 = _____   b. 418 + 10 = _____

c. 103 + 10 = _____   d. 10 + 412 = _____

Part 7

a. 10 x 6 = _____   b. 4 x 3 = _____

c. 2 x 5 = _____   d. 5 x 4 = _____

116 Lesson 44   Connecting Math Concepts

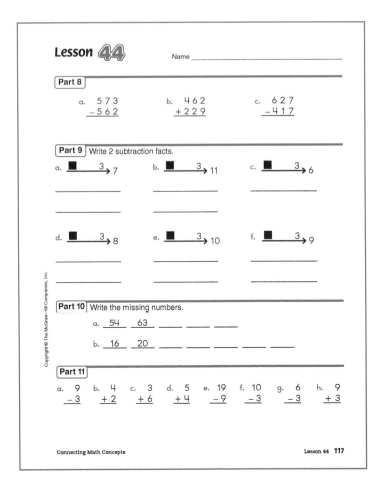

Lesson 44

Name _____

Part 8

a. 573
 − 562

b. 462
 + 229

c. 627
 − 417

Part 9 Write 2 subtraction facts.

a. ■ —3→ 7

b. ■ —3→ 11

c. ■ —3→ 6

_____ _____ _____

_____ _____ _____

d. ■ —3→ 8

e. ■ —3→ 10

f. ■ —3→ 9

_____ _____ _____

_____ _____ _____

Part 10 Write the missing numbers.

a. 54 63 ____ ____ ____

b. 16 20 ____ ____ ____ ____ ____

Part 11

a. 9
 − 3

b. 4
 + 2

c. 3
 + 6

d. 5
 + 4

e. 19
 − 9

f. 10
 − 3

g. 6
 − 3

h. 9
 + 3

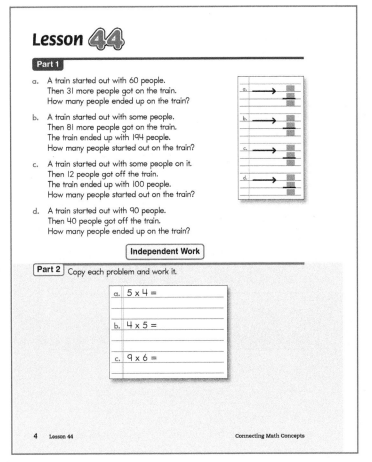

Lesson 44

Part 1

a. A train started out with 60 people.
Then 31 more people got on the train.
How many people ended up on the train?

b. A train started out with some people.
Then 81 more people got on the train.
The train ended up with 194 people.
How many people started out on the train?

c. A train started out with some people on it.
Then 12 people got off the train.
The train ended up with 100 people.
How many people started out on the train?

d. A train started out with 90 people.
Then 40 people got off the train.
How many people ended up on the train?

Independent Work

Part 2 Copy each problem and work it.

a. 5 × 4 =

b. 4 × 5 =

c. 9 × 6 =

Lesson

EXERCISE 1: DATA TABLES
COLUMNS AND ROWS

a. Open your workbook to Lesson 102 and find part 1. ✔
(Teacher reference:)

	A	B	C
☆	8	5	11
△	2	10	6
☽	9	0	7

a. Circle the largest number in the △ row.
b. Circle the smallest number in the ☆ row.
c. Cross out the smallest number in column C.
d. Cross out the largest number in column A.

b. I'll read item A: Circle the largest number in the triangle row.
- What are you going to do with the largest number in the triangle row? (Signal.) *Circle it.*
- Touch the triangle row. ✔
- Read the numbers in the triangle row. (Signal.) *2, 10, 6.*
- Circle the largest number.
(Observe students and give feedback.)
(Display:)   [102:1A]

	A	B	C
☆	8	5	11
△	2	⑩	6
☽	9	0	7

Here's what you should have.
- The numbers are 2, 10, and 6.
  Which number did you circle? (Signal.) *10.*
c. I'll read item B: Circle the smallest number in the star row.
- What are you going to do with the smallest number in the star row? (Signal.) *Circle it.*
- Find the star row and circle the smallest number.
(Observe students and give feedback.)
(Add to show:)   [102:1B]

	A	B	C
☆	8	⑤	11
△	2	⑩	6
☽	9	0	7

Here's what you should have.

- The numbers are 8, 5, and 11. Which number did you circle? (Signal.) *5.*
d. I'll read item C: Cross out the smallest number in column C.
- What are you going to do with the smallest number in column C? (Signal.) *Cross it out.*
- Touch column C. ✔
- Say the numbers in column C. (Signal.) *11, 6, 7.*
- Cross out the smallest number.
(Observe students and give feedback.)
(Add to show:)   [102:1C]

	A	B	C
☆	8	⑤	11
△	2	⑩	✕
☽	9	0	7

Here's what you should have.
- The numbers are 11, 6, and 7. Which number did you cross out? (Signal.) *6.*
e. I'll read item D: Cross out the largest number in column A.
- What are you going to do with the largest number in column A? (Signal.) *Cross it out.*
- Find column A and cross out the largest number.
(Observe students and give feedback.)
(Add to show:)   [102:1D]

	A	B	C
☆	8	⑤	11
△	2	⑩	✕
☽	✕	0	7

Here's what you should have.
- The numbers are 8, 2, and 9. Which number did you cross out? (Signal.) *9.*

EXERCISE 2: EQUIVALENT UNITS

>, <, =   **REMEDY**

a. You learned that there are 60 seconds in a minute.

• How many seconds are in a minute? (Signal.) *60.*

• Yes, 60 seconds equals 1 minute.
  Say the rule. (Signal.) *60 seconds equals 1 minute.*

b. You learned that there are 100 centimeters in a meter.

• How many centimeters are in a meter? (Signal.) *100.*
  Yes, 100 centimeters equals 1 meter.

• Say the rule. (Signal.) *100 centimeters equals 1 meter.*

c. Say the rule about seconds and minutes. (Signal.) *60 seconds equals 1 minute.*

• Say the rule about centimeters and meters. (Signal.) *100 centimeters equals 1 meter.*
  (Repeat until firm.)

━━━━━━ **WORKBOOK PRACTICE** ━━━━━━

a. Find part 2 in your workbook. ✔
  (Teacher reference:)   **R** | **Part H**

a. 25 seconds ___ 1 minute   e. 110 centimeters ___ 1 meter

b. 1 meter ___ 65 centimeters   f. 1 yard ___ 3 feet

c. 1 week ___ 9 days   g. 1 gallon ___ 3 quarts

d. 12 months ___ 1 year   h. 95 cents ___ 1 dollar

  You're going to write the missing sign for each item.

b. Item A: 25 seconds, 1 minute.

• Say the rule for seconds and minutes. (Signal.) *60 seconds equals 1 minute.*

• Is 25 seconds more than, less than, or equal to 1 minute? (Signal.) *Less than 1 minute.*

c. Item B: 1 meter, 65 centimeters.

• Say the rule for centimeters and meters. (Signal.) *100 centimeters equals 1 meter.*

• Is 65 centimeters more than, less than, or equal to 1 meter? (Signal.) *Less than 1 meter.*

d. Your turn: Write the missing sign in each item.

• Pencils down when you're finished.
  (Observe students and give feedback.)

e. Check your work.
  Here's what you should have for each item.
  (Display:)   [102:2A]

> a. 25 seconds < 1 minute
> b. 1 meter   > 65 centimeters
> c. 1 week   < 9 days
> d. 12 months = 1 year

f. (Point to **A**.) Everybody, read statement A. (Signal.) *25 seconds is less than 1 minute.*

• How many seconds are in a minute? (Signal.) *60.*

g. (Point to **B**.) Read statement B. (Signal.) *1 meter is more than 65 centimeters.*

• How many centimeters are in a meter? (Signal.) *100.*

h. (Point to **C**.) Read statement C. (Signal.) *1 week is less than 9 days.*

• How many days are in a week? (Signal.) *7.*

i. (Point to **D**.) Read statement D. (Signal.) *12 months equals 1 year.*
  (Display:)   [102:2B]

> e. 110 centimeters > 1 meter
> f. 1 yard   = 3 feet
> g. 1 gallon > 3 quarts
> h. 95 cents < 1 dollar

j. (Point to **E**.) Read statement E. (Signal.) *110 centimeters is more than 1 meter.*

• How many centimeters are in a meter? (Signal.) *100.*

k. (Point to **F**.) Read statement F. (Signal.) *1 yard equals 3 feet.*

l. (Point to **G**.) Read statement G. (Signal.) *1 gallon is more than 3 quarts.*

• How many quarts are in a gallon? (Signal.) *4.*

m. (Point to **H**.) Read statement H. (Signal.) *95 cents is less than 1 dollar.*

• How many cents are in a dollar? (Signal.) *100.*

EXERCISE 3: MONEY EQUIVALENCE
CENTS/DOLLARS AND CENTS

a. Find part 3 in your workbook. ✔
(Teacher reference:)

a. 908 cents =   b. 340 cents =

c. 779 cents =   d. 215 cents =

You're going to write the dollars and cents number for each cents number.

b. Problem A: Write the dollars and cents number for 908 cents. ✔
(Display:)   [102:3A]

> **a. 908 cents = $9.08**

Here's what you should have.

• Problem B: Write the dollars and cents number for 340 cents. ✔
(Display:)   [102:3B]

> **b. 340 cents = $3.40**

Here's what you should have.

• Problem C: Write the dollars and cents number for 779 cents. ✔
(Display:)   [102:3C]

> **c. 779 cents = $7.79**

Here's what you should have.

• Problem D: Write the dollars and cents number for 215 cents. ✔
(Display:)   [102:3D]

> **d. 215 cents = $2.15**

Here's what you should have.

EXERCISE 4: ADDITION FACTS
SMALL NUMBER OF 8

a. (Hand out lined paper.)
• Pencils down. ✔
(Display:)   [102:4A]

$$\underset{\longrightarrow}{8 \quad 3} \qquad \underset{\longrightarrow}{8 \quad 4}12 \qquad \underset{\longrightarrow}{8 \quad 5}13$$

b. (Point to $\underset{\longrightarrow}{8 \quad 3}$.) What are the small numbers in this family? (Signal.) *8 and 3.*
• What's the big number? (Signal.) *11.*
• Say the addition fact that starts with 8. (Signal.) *8 + 3 = 11.*
• Say the other addition fact. (Signal.) *3 + 8 = 11.*
c. (Point to $\underset{\longrightarrow}{8 \quad 4}$12 and $\underset{\longrightarrow}{8 \quad 5}$13.) Here are two new number families.

⌐ d. (Point to $\underset{\longrightarrow}{8 \quad 4}$12.) What are the small numbers in this family? (Signal.) *8 and 4.*
  • Say the addition fact that starts with 8. (Signal.) *8 + 4 = 12.*
  • Say the other addition fact. (Signal.) *4 + 8 = 12.*
  └ (Repeat until firm.)
⌐ e. (Point to $\underset{\longrightarrow}{8 \quad 5}$13.) Say the addition fact that starts with 8. (Signal.) *8 + 5 = 13.*
  • Say the other addition fact. (Signal.) *5 + 8 = 13.*
  └ (Repeat until firm.)
f. (Change to show:)   [102:4B]

$$\underset{\longrightarrow}{8 \quad 3}, \qquad \underset{\longrightarrow}{8 \quad 4}, \qquad \underset{\longrightarrow}{8 \quad 5},$$

Here are the families without the big numbers.
⌐ g. (Point to $\underset{\longrightarrow}{8 \quad 3}$.) Say the addition fact that starts with 8. (Signal.) *8 + 3 = 11.*
  • (Point to $\underset{\longrightarrow}{8 \quad 4}$.) Say the addition fact that starts with 8. (Signal.) *8 + 4 = 12.*
  • (Point to $\underset{\longrightarrow}{8 \quad 5}$.) Say the addition fact that starts with 8. (Signal.) *8 + 5 = 13.*
  └ (Repeat until firm.)
⌐ h. (Point to $\underset{\longrightarrow}{8 \quad 3}$.) What does 3 plus 8 equal? (Signal.) *11.*
  Say the fact. (Signal.) *3 + 8 = 11.*
  • (Point to $\underset{\longrightarrow}{8 \quad 4}$.) Say the fact that starts with 4. (Signal.) *4 + 8 = 12.*
  • (Point to $\underset{\longrightarrow}{8 \quad 5}$.) Say the fact that starts with 5. (Signal.) *5 + 8 = 13.*
  └ (Repeat until firm.)

i. (Display:)   [102:4C]

$$8 + 4$$
$$8 + 5$$
$$8 + 3$$
$$5 + 8$$
$$4 + 8$$
$$3 + 8$$

Here are problems that have a small number of 8.
- (Point to **8 + 4**.) Read the problem. (Signal.) *8 + 4.*
  What's the answer? (Signal.) *12.*
- (Point to **8 + 5**.) Read the problem. (Signal.) *8 + 5.*
  What's the answer? (Signal.) *13.*
- (Point to **8 + 3**.) Read the problem. (Signal.) *8 + 3.*
  What's the answer? (Signal.) *11.*
- (Point to **5 + 8**.) Read the problem. (Signal.) *5 + 8.*
  What's the answer? (Signal.) *13.*
- (Point to **4 + 8**.) Read the problem. (Signal.) *4 + 8.*
  What's the answer? (Signal.) *12.*
- (Point to **3 + 8**.) Read the problem. (Signal.) *3 + 8.*
  What's the answer? (Signal.) *11.*
  (Repeat until firm.)
j. (Change to show:)   [102:4D]

a. $5 + 8$
b. $8 + 4$
c. $3 + 8$
d. $8 + 5$
e. $8 + 3$
f. $4 + 8$

Here are the problems in a different order.
k. Write A through F on your lined paper. ✔
- Now write the answer to each problem.
  (Observe students and give feedback.)
l. Check your work.
- Problem A: What's 5 plus 8? (Signal.) *13.*
  Say the fact for 5 plus 8. (Signal.) *5 + 8 = 13.*

- Problem B: What's 8 plus 4? (Signal.) *12.*
  Say the fact for 8 plus 4. (Signal.) *8 + 4 = 12.*
- Problem C: What's 3 plus 8? (Signal.) *11.*
  Say the fact for 3 plus 8. (Signal.) *3 + 8 = 11.*
- Problem D: What's 8 plus 5? (Signal.) *13.*
  Say the fact for 8 plus 5. (Signal.) *8 + 5 = 13.*
- Problem E: What's 8 plus 3? (Signal.) *11.*
  Say the fact for 8 plus 3. (Signal.) *8 + 3 = 11.*
- Problem F: What's 4 plus 8? (Signal.) *12.*
  Say the fact for 4 plus 8. (Signal.) *4 + 8 = 12.*
m. (Display:)   [102:4E]

Remember the two new families.
- (Point to $\underset{\longrightarrow}{8 \quad 4}$.) What are the small numbers in this family? (Signal.) *8 and 4.*
  What's the big number? (Signal.) *12.*
- (Point to $\underset{\longrightarrow}{8 \quad 5}$.) What are the small numbers in this family? (Signal.) *8 and 5.*
  What's the big number? (Signal.) *13.*

EXERCISE 5: MONEY WORD PROBLEMS
TOTAL PURCHASE

a. (Hand out textbooks.)
- Open your textbook to Lesson 102 and find part 1. ✔
  (Teacher reference:)

The pictures show price tags for items in a store.
- Touch the boots. ✔
  What's the price of the boots? (Signal.) *$31.60.*
- Touch the coat. ✔
  What's the price of the coat? (Signal.) *$44.25.*
- Touch the shirt. ✔
  What's the price of the shirt? (Signal.) *$12.95.*
- Touch the gloves. ✔
  What's the price of the gloves? (Signal.) *$10.00.*
  (Repeat until firm.)
b. Listen: Raise your hand when you know how much money you need to buy the shirt. ✔
- How much money would you need? (Signal.) *$12.95.*

c. Raise your hand when you know how much you would need to buy the boots. ✔
• How much money would you need? (Signal.) *$31.60.*

d. Raise your hand when you know how much money you would need to buy the coat. ✔
• How much money would you need? (Signal.) *$44.25.*

e. Raise your hand when you know how much money you would need to buy the gloves. ✔
• How much money would you need? (Signal.) *$10.00.*

f. I'll read problem A: You want to buy the boots and the gloves. How much money do you need?
• You're going to figure out how much money you need to buy the boots and the gloves. Work problem A.
(Observe students and give feedback.)

g. Check your work.
(Display:)   [102:5A]

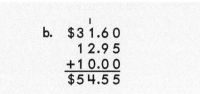

Here's what you should have.
$31.60 + $10.00 = $41.60. So you would need $41.60 to buy the boots and the gloves.

h. I'll read problem B: You want to buy the coat and the shirt. How much money do you need?
• You're going to figure out how much money you need to buy the coat and the shirt. Work problem B.
(Observe students and give feedback.)

i. Check your work.
(Display:)   [102:5B]

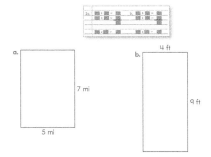

Here's what you should have.
$44.25 + $12.95 = $57.20. So you would need $57.20 to buy the coat and the shirt.

j. I'll read problem C: You want to buy the boots, the shirt, and the gloves. How much money do you need?
• You're going to figure out how much money you need to buy the boots, the shirt, and the gloves. Work problem C.
(Observe students and give feedback.)

k. Check your work.
(Display:)   [102:5C]

b. $31.60
     12.95
   +10.00
   $54.55

Here's what you should have.
$31.60 + $12.95 + $10.00 = $54.55. So you would need $54.55 to buy the boots, the shirt, and the gloves.

EXERCISE 6: AREA AND PERIMETER
OF RECTANGLES   **REMEDY**

a. Find part 2 in your textbook. ✔
(Teacher reference:)

You're going to find the area and the perimeter of each rectangle.

b. Remember, when you find the **area,** the answer has **square** units. When you find the **perimeter,** the answer does **not** have square units.

c. Touch rectangle A. ✔
• Listen: When you work the problem for the **perimeter,** what's the unit name in the answer? (Signal.) *Miles.*
• When you find the **area,** what's the unit name in the answer? (Signal.) *Square miles.*
(Repeat until firm.)

d. Find the perimeter of rectangle A. Then stop.
(Observe students and give feedback.)
• Everybody, read the equation for the top and bottom sides. (Signal.) *5 + 5 = 10.*
• Read the equation for the up-and-down sides. (Signal.) *7 + 7 = 14.*
• What's the perimeter? (Signal.) *24 miles.*
Yes, 24 miles.

e. Find the area of rectangle A.
(Observe students and give feedback.)
- You worked the problem 5 times 7. What's the area? (Signal.) *35 square miles.*
Yes, 35 **square** miles.

f. Touch rectangle B. ✔
- Listen: When you work the problem for the perimeter, what's the unit name in the answer? (Signal.) *Feet.*
- When you find the area, what's the unit name in the answer? (Signal.) *Square feet.*
(Repeat until firm.)

g. Find the perimeter of rectangle B. Then stop.
(Observe students and give feedback.)
- Everybody, read the equation for the top and bottom sides. (Signal.) *4 + 4 = 8.*
- Read the equation for the up-and-down sides. (Signal.) *9 + 9 = 18.*
- What's the perimeter? (Signal.) *26 feet.*
Yes, 26 feet.

h. Find the area of rectangle B.
(Observe students and give feedback.)
- You worked the problem 4 times 9. What's the area? (Signal.) *36 square feet.*
Yes, 36 square feet.

EXERCISE 7: INDEPENDENT WORK
MULTIPLICATION

a. Find part 3 in your textbook. ✔
(Teacher reference:)

a. $5 \times \blacksquare = 15$ d. $2 \times 8 = \blacksquare$
b. $10 \times \blacksquare = 40$ e. $2 \times \blacksquare = 12$
c. $9 \times 6 = \blacksquare$ f. $4 \times 5 = \blacksquare$

3a.	d.
b.	e.
c.	f.

My turn to read problems A and B.
- Problem A: 5 times what number equals 15.
- Problem B: 10 times what number equals 40.

b. Your turn: Read problem A. (Signal.) *5 x what number = 15.*
- Read problem B. (Signal.) *10 x what number = 40.*
- Read problem C. (Signal.) *9 x 6 = what number.*

c. You'll copy all the problems and work them when you do your independent work.

Assign Independent Work, Textbook parts 3–7 and Workbook parts 4–6.

Lesson 102

Name _____

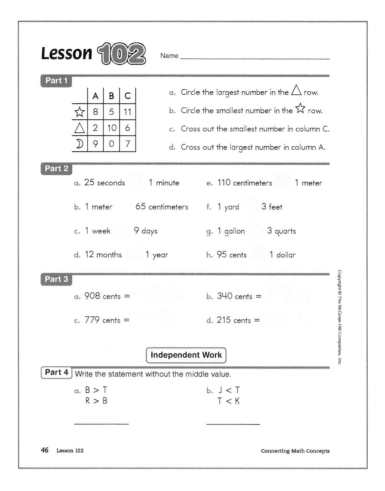

Part 1

	A	B	C
☆	8	5	11
△	2	10	6
☽	9	0	7

a. Circle the largest number in the △ row.

b. Circle the smallest number in the ☆ row.

c. Cross out the smallest number in column C.

d. Cross out the largest number in column A.

Part 2

a. 25 seconds   1 minute

b. 1 meter   65 centimeters

c. 1 week   9 days

d. 12 months   1 year

e. 110 centimeters   1 meter

f. 1 yard   3 feet

g. 1 gallon   3 quarts

h. 95 cents   1 dollar

Part 3

a. 908 cents =

b. 340 cents =

c. 779 cents =

d. 215 cents =

Independent Work

Part 4 Write the statement without the middle value.

a. B > T
   R > B

b. J < T
   T < K

_____

_____

Lesson 102

Name _____

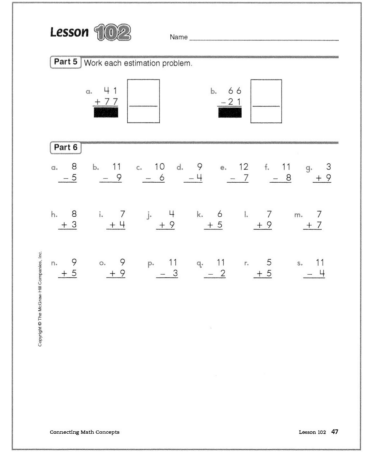

Part 5 Work each estimation problem.

a.   4 1
   + 7 7   ▢

b.   6 6
   − 2 1   ▢

Part 6

a.   8   b. 11   c. 10   d.   9   e. 12   f. 11   g.   3
   − 5   − 9   − 6   − 4   − 7   − 8   + 9

h.   8   i.   7   j.   4   k.   6   l.   7   m.   7
   + 3   + 4   + 9   + 5   + 9   + 7

n.   9   o.   9   p. 11   q. 11   r.   5   s. 11
   + 5   + 9   − 3   − 2   + 5   − 4

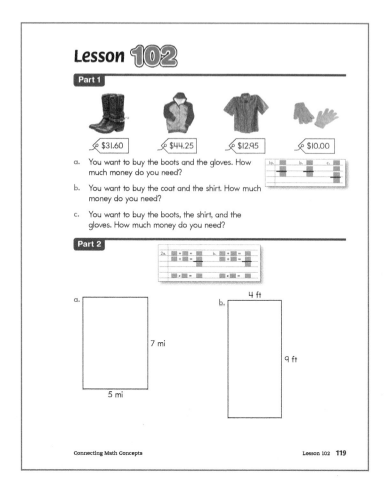

Lesson 102

Part 1

$31.60 $44.25 $12.95 $10.00

a. You want to buy the boots and the gloves. How much money do you need?

b. You want to buy the coat and the shirt. How much money do you need?

c. You want to buy the boots, the shirt, and the gloves. How much money do you need?

Part 2

a. (rectangle: 7 mi by 5 mi)

b. (rectangle: 4 ft by 9 ft)

Connecting Math Concepts   Lesson 102   **119**

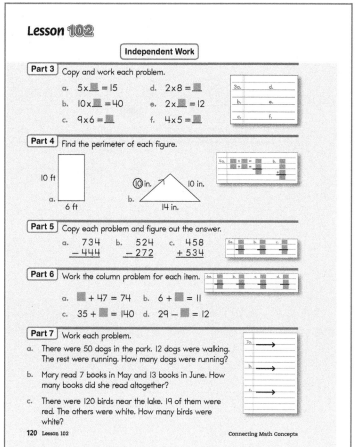

Lesson 102

Independent Work

Part 3 Copy and work each problem.

a. $5 \times \blacksquare = 15$ d. $2 \times 8 = \blacksquare$

b. $10 \times \blacksquare = 40$ e. $2 \times \blacksquare = 12$

c. $9 \times 6 = \blacksquare$ f. $4 \times 5 = \blacksquare$

Part 4 Find the perimeter of each figure.

a. (rectangle: 10 ft by 6 ft)

b. (triangle: 10 in., 10 in., 14 in.)

Part 5 Copy each problem and figure out the answer.

a. $\begin{array}{r} 734 \\ -444 \end{array}$ b. $\begin{array}{r} 524 \\ -272 \end{array}$ c. $\begin{array}{r} 458 \\ +534 \end{array}$

Part 6 Work the column problem for each item.

a. $\blacksquare + 47 = 74$ b. $6 + \blacksquare = 11$

c. $35 + \blacksquare = 140$ d. $29 - \blacksquare = 12$

Part 7 Work each problem.

a. There were 50 dogs in the park. 12 dogs were walking. The rest were running. How many dogs were running?

b. Mary read 7 books in May and 13 books in June. How many books did she read altogether?

c. There were 120 birds near the lake. 19 of them were red. The others were white. How many birds were white?

120 Lesson 102   Connecting Math Concepts
